UNIVERSITY OPTICS

UNIVERSITY OPTICS

Volume 1

D. W. TENQUIST, M.SC.
R. M. WHITTLE, B.SC., PH.D.
J. YARWOOD, M.SC., F.INST.P.

LONDON   ILIFFE BOOKS LTD

ILIFFE BOOKS LTD
42, RUSSELL SQUARE
LONDON, W.C.1

First published in 1969

592   05054   8

Printed in Hungary

Contents

Preface

This text is intended to cover the requirements in the subject of optics for the student preparing for Part 1 of an Honours Degree in Physics, a General Degree in Science, or Ancillary Physics to an Honours Degree in Chemistry or other main discipline. It is also suitable for students reading for the Higher National Certificate or Diploma in Physics, or the Graduateship Examination of The Institute of Physics and The Physical Society.

It is assumed that the reader has a preliminary knowledge of physics and mathematics at the standard of Advanced Level General Certificate of Education or the equivalent.

It is appreciated that the tendency exists in some syllabuses to integrate the study of optics within an account of vibrations, waves and radiations. It is regarded as convenient and useful, nevertheless, that the present text be concerned chiefly with the visible, infra-red and ultra-violet regions of the electromagnetic spectrum. To do justice to both the theoretical and experimental aspects involved would result in an unduly lengthy work if other than appropriate mention were made of other regions of the electromagnetic spectrum and of the wave motion aspects of the mechanics of solids and fluids, acoustics and particles.

The text is divided into two volumes, which can be studied independently of one another. In the selection of material to be included within these two books, the choice has been guided primarily by the knowledge to be expected of an undergraduate student who is concurrently studying mathematics and other branches of physics. The first volume can be read with a mathematical equipment little beyond that acquired at school; the second one demands an appreciation of the mathematics and other branches of physics which are usually included in the first year at university or college.

At the end of each chapter is included a number of exercises, mostly taken from the degree examination papers of various universities; acknowledgements to these sources are made as appropriate in the text.

The authors express their sincere thanks to Mrs Elizabeth Bangham and Mrs Angela Williams for their typing of the manuscript.

D. W. T.

LONDON   R. M. W.

1969   J. Y.

CHAPTER 1

Lenses and Systems of Lenses

The theory and practice of geometrical optics is based on the rectilinear propagation of radiation in a uniform medium of constant refractive index. A luminous point source emits radiant energy. Within a homogeneous medium free from obstacles, this energy travels equally in all directions away from the source to form a spherical wavefront. Any radius of this sphere is a *light ray*. If a ray passes from one medium into another of different refractive index, it is deviated. A bundle of rays proceeding from a point source forms a cone referred to as a divergent *pencil of light*. Conversely, a bundle of rays may converge, or tend to converge, in a cone towards a point to form a convergent pencil of light.

1.1 An Assembly of Coaxial Spherical Refracting Surfaces

In a simple spherical lens, the thickness of the glass (or other medium) along the axis of symmetry is small compared with the distances from the centre of the lens to the object and to the image. The theoretical treatment of a simple spherical lens can be extended to a number of lenses arranged coaxially, the common axis of symmetry being the *optical axis*, and also to a thick lens, which is one of significant axial thickness compared with the object and the image distances.

In the treatment of a system of coaxial lenses, the cones of light are restricted so that the semi-angle θ (the angle between the outermost ray of the cone and its axis of symmetry) is small enough to enable first-order approximations of trigonometrical ratios to be used: i.e. $\theta = \sin \theta = \tan \theta$ and $\cos \theta = 1$. Furthermore, the axis of the cone must be parallel, or nearly parallel, to the optical axis of the system and separated from this axis by distances which are short compared with the radius of the aperture of any lens in the system. These limitations define a *paraxial* cone of light. Under these conditions, it can be assumed that all rays from a given object point pass through a given image point after traversing a system of coaxial refracting surfaces; the treatment of such a system is greatly simplified by consideration of the nature and position of an imaginary thin lens that would give the same object and image positions. To specify this equivalent lens, the *cardinal points* of the system need to be defined.

1.2 Cardinal Points of a System of Coaxial Spherical Refracting Surfaces

There are six cardinal points in the form of three pairs of conjugate points known as the *focal points*, the *principal points* and the *nodal points*.

The focal points of an optical system are two points on the optical axis conjugate to points at infinity. The region in which an object may be situated and from which light is incident upon the optical system is called the *object space;* in this text, the object is usually taken to be on the left-hand side of the optical system— this choice of left being purely arbitrary. The region into which the light emerges, or appears to emerge, after refraction is the *image space;* this is where the image (which may be real or virtual) lies and it may be, but is not necessarily, to the right of the system. The object and image spaces may overlap.

Ignoring the actual paths of the rays of light within the optical system, suppose that the ray *AB* is incident parallel to the optical axis in the object space (Fig. 1.1). So that a diagram can be constructed, the optical system is presumed to be converging (a diverging system is considered in Section 1.3). After refractions in the optical system, the ray *AB* will emerge along *CF'* to intersect the optical axis in the image space at point *F'*. The point *F'* is the *second focal point* of the system (a dash, as in *F'*, is used for all image space cardinal points). A plane drawn through *F'* perpendicular to the optical axis is the *second focal plane* of the system. Why 'second'? This is simply in accordance with the convention that the object space is 'first' and the image space is 'second'—following the order in which they are traversed by the rays of light. *F'* is thus conjugate to a point at infinity in the object space. Conversely, if the initial ray parallel to the optical axis is incident from the right along $A'B'$ in the image space, the conjugate point F will be the *first focal point* in the object space, and the *first focal plane* will be perpendicular to the optical axis and will include F.

The principal points of an optical system are two conjugate points on the optical axis such that principal planes drawn perpendicular to the optical axis through these points are planes of unit magnification. Such planes are obviously convenient reference positions because an object placed at one of them will give rise to an image of the same lateral dimensions at the other. It may well happen in practice that a principal plane is actually within the glass of the lens, and thus a real object cannot be located in it; but this does not invalidate the usefulness of the concept.

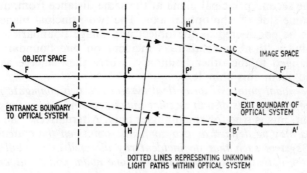

Fig. 1.1. Focal points and principal points of a converging optical system, the rays of light being paraxial

In Fig. 1.1, let the paraxial ray AB parallel to the optical axis be produced to cut $F'C$ produced at H'. Similarly, by proceeding in the opposite direction along $A'B'$ in the image space, the point H can be found. It is simple to show that these points H and H' lie in the principal planes of the system, which are perpendicular to the optical axis at the first and second principal points P and P' respectively. Thus, in Fig. 1.2, let PQ be a small object perpendicular to the optical axis placed at the first principal point P. To find the image of Q, the well-known construction is followed: first, the incident ray FQ (which, by definition, will emerge along $Q'Y$ parallel to the optical axis) is drawn and, second, the incident ray XQ parallel to the optical axis (which must emerge through F' and appear to come from Q'). Clearly $P'Q'$ is then the image of PQ and $P'Q' = PQ$.

It follows from the above analysis that the position of any image can be found by drawing appropriate rays incident from the

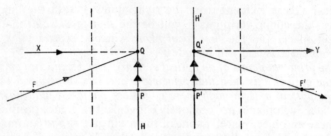

Fig. 1.2. Showing that an object PQ at one principal plane gives rise to an image of the same lateral dimensions at P'Q' in the other principal plane

object on to the first principal plane, each emerging from a point on the second principal plane at the same distance from, and on the same side of, the optical axis. The two principal planes can hence be considered as the object and image boundaries of an imaginary 'thin' lens; all rays incident on one boundary pass unchanged to the other boundary, where they emerge as they would from this thin lens.

The nodal points of an optical system are two conjugate points on the optical axis of unit angular magnification. Again, for a real system, this is true for small angles only. Alternatively, the *nodal points may be defined as two conjugate points on the optical axis of a system such that an incident ray directed at a small angle θ to the optical axis to pass through one nodal point will emerge as if from the other nodal point along a line parallel to the incident ray and so making the same angle θ with the optical axis.*

In Fig. 1.3, the nodal points are at N and N'. Consider an incident ray AB at an angle α to the optical axis. When produced, this ray intersects the axis at the first nodal point N; the conjugate ray B'A' will be along a line such that N'B'A' makes the same angle α with the optical axis, N' being the conjugate second

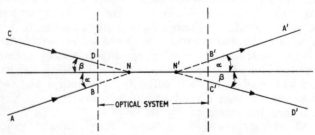

Fig. 1.3. Nodal points of an optical system

nodal point. A second incident ray CD produced to N and making an angle β with the optical axis will emerge along $C'D'$, where $N'C'D'$ is also at angle β to the optical axis. The two definitions given are thus equivalent because the angles defined by the traces CNA and $C'N'A'$ of the incident and emergent cones are equal.

The nodal points correspond to the 'optical centre' of the equivalent thin lens. In the optical system, two principal points are needed, as against one in the simple thin lens; likewise two nodal points are required instead of one optical centre.

1.3 Further Consideration of the Analogy between the Optical System and a Thin Lens

The cardinal points of the general coaxial optical system are shown in Fig. 1.4. The distances $PF = f$ and $P'F' = f'$ are defined as the first and second focal lengths of the system by obvious analogy with the thin lens. Note that f and f' are not necessarily equal numerically. As will be shown later, f *is* equal to f' numerically provided that the media of the object and image spaces have the same refractive index (the frequently encountered case being when they are both air); this is, of course, also true for a thin lens.

To find, by ray-tracing, the position of the image of a given object, the procedure for the optical system is the same as for the thin lens; this can be seen from Fig. 1.5, where the thin lens may be regarded as a 'coalescence' of the two principal planes. In each part of Fig. 1.5, only two of the three constructional rays shown are needed to locate the image.

Sometimes, the principal points are 'crossed', i.e. the separation between P' and F is less than that between P and F [Fig. 1.6(a)]. In a diverging system, the focal points F and F' will be between

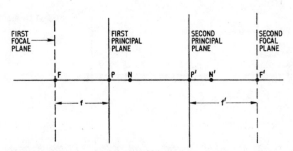

Fig. 1.4. Defining the focal lengths of a general optical system

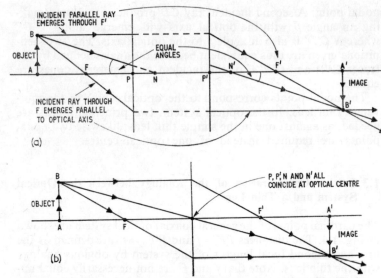

Fig. 1.5. Ray-tracing procedure to establish the position of an image of a given
object: (a) the general optical system; (b) the thin lens

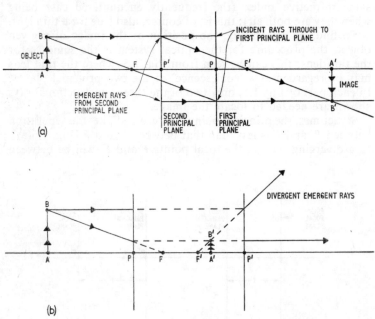

Fig. 1.6. (a) The 'crossed' general optical system; (b) the divergent general
optical system

P and P' [Fig. 1.6(b)], instead of outside them as in the converging system. Nevertheless, the same rules for locating the image of an object by ray-tracing still apply. Thus, the first principal plane is the one to which all incident rays from the object are taken, even if it is beyond the second principal plane, as in Fig. 1.6(a).

Any ray may be traced through the optical system, not only the particular parallel rays and rays through the foci so far discussed. To illustrate this, consider a converging system comprising two separated thin lenses L_1 and L_2 and having cardinal

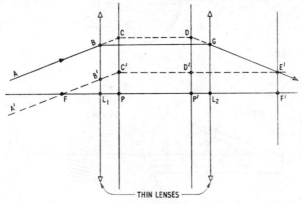

Fig. 1.7. Ray-tracing through a system comprising two separated thin lenses

points F, F', P and P' (Fig. 1.7). (Note the symbols used for thin lenses in Fig. 1.7.) AB is a ray incident on L_1 in any direction. A ray $A'B'$, which passes through F and is parallel to AB, is traced. From knowledge of focal and principal points, $A'B'$ can be produced to C' within the first principal plane, pass to D' in the second principal plane (with $C'P = D'P'$) and emerge parallel to the optical axis; at E', it cuts the second focal plane of the system. As ray AB is initially parallel to $A'B'$, it also must pass through E' on emergence from the system. Hence the path of AB must be $ABCDE'$ via the principal planes, and $ABGE'$ through the two thin lenses.

1.4 Spherical Refracting Surface Separating Two Media

Let r be the radius of curvature of a spherical surface which has a centre of curvature at C and separates two media 1 and 2 of refractive indices n and n' respectively (Fig. 1.8). P is a point object within the medium 1 at a distance x from the pole of the refracting surface at O; Q is the conjugate image point where $OQ = x'$. The refracting surface is either convex towards the object point [Fig. 1.8(a)], or it is concave towards the object point [Fig. 1.8(b)].

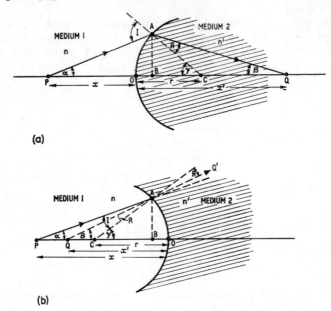

(a)

(b)

Fig. 1.8. Spherical refracting surface separating two media: (a) surface convex towards the object point; (b) surface concave towards the object point

In Fig. 1.8(a), the incident ray PA at an angle of incidence I to the normal CA is refracted along AQ, where the angle of refraction $CAQ = R$. From Snell's law,

$$\frac{\sin I}{\sin R} = \frac{n'}{n}$$

To a first approximation, $\sin I = I$ and $\sin R = R$ for paraxial rays; therefore,

$$nI = n'R \tag{1.1}$$

Noting the angles α, β and γ shown in Fig. 1.8(a), and proceeding geometrically *disregarding any optical sign convention:*

$$R = \gamma - \beta$$

and

$$I = \alpha + \gamma$$

Substitution of these expressions for R and I into Equation 1.1 gives:

$$n(\alpha + \gamma) = n'(\gamma - \beta) \tag{1.2}$$

If AB is constructed perpendicular to the optical axis, it follows that, for paraxial rays:

$$\alpha = \frac{AB}{x}, \quad \beta = \frac{AB}{x'} \quad \text{and} \quad \gamma = \frac{AB}{r}$$

Substitution in Equation 1.2 gives:

$$n\left(\frac{AB}{x} + \frac{AB}{r}\right) = n'\left(\frac{AB}{r} - \frac{AB}{x'}\right)$$

Therefore,

$$\frac{n'}{x'} + \frac{n}{x} = \frac{n'-n}{r} \tag{1.3}$$

If the refracting surface is concave towards the object point [Fig. 1.8(b)], then as before,

$$nI = n'R$$

and

$$R = \gamma - \beta$$

But now,

$$I = \gamma - \alpha$$

So

$$n(\gamma - \alpha) = n'(\gamma - \beta)$$

Therefore,

$$n\left(\frac{AB}{r} - \frac{AB}{x}\right) = n'\left(\frac{AB}{r} - \frac{AB}{x'}\right)$$

and

$$\frac{n'}{x'} - \frac{n}{x} = \frac{n'-n}{r} \tag{1.4}$$

Note the difference between Equations 1.3 and 1.4.

1.5 The Concept of Vergence

The trace of a cone of rays through several refracting surfaces in an optical system can be calculated by repeated use of Equations 1.3 and 1.4. Although this procedure is often used, it can be very tedious; the idea of vergence provides a useful alternative.

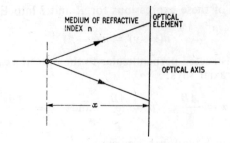

Fig. 1.9. The vergence of a cone of rays

The term 'vergence' is used in relation to either convergence or divergence. The vergence of a cone of rays is defined as n/x, where n is the refractive index of the uniform medium in which the cone is situated, and x is the distance along the optical axis from the apex of this cone to the surface of an optical element (lens or mirror). If the cone is converging, the vergence is said to be positive, i.e. n/x is positive; a diverging cone (Fig. 1.9). will have a negative vergence. If vergence is denoted by V,

$$V = \frac{n}{x} \tag{1.5}$$

The *power of an optical system* is defined as the difference between the vergence in the image space and the vergence in the object space. Thus, in Fig. 1.10 which represents a converging system, the power F is given by:

$$F = V' - V \tag{1.6}$$

If a spherical refracting surface separates two media as in Fig. 1.8, the vergence V of the incident cone from the object point

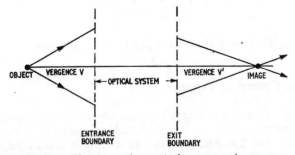

Fig. 1.10. The power of an optical system, and vergence

P is $-n/x$ since the cone is divergent. But in Fig. 1.8(a), $V' = n'/x'$; whereas in Fig. 1.8(b), $V' = -n'/x'$. Hence, in Fig. 1.8(a),

$$F = V' - V = \frac{n'}{x'} + \frac{n}{x} = \frac{n' - n}{r}$$

from Equation 1.3; similarly for Fig. 1.8(b), using Equation 1.4,

$$F = V' - V = -\frac{n'}{x'} + \frac{n}{x} = \frac{n - n'}{r}$$

The power of a spherical refracting surface is taken to be positive if the surface (or, generally, the optical element) makes the incident beam more convergent (or less divergent), and negative if the incident beam is made more divergent (or less convergent).

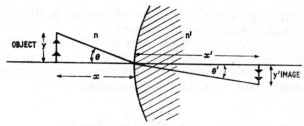

Fig. 1.11. The linear magnification produced by a spherical refracting surface

If it is remembered that the optically denser medium (i.e. the medium with the larger refractive index n) bends an incident ray towards the normal to the refracting surface, the sign of F presents no problem. Thus, in Fig. 1.8(a), the incident cone is converged by the surface, $n' > n$, and F is positive; in Fig. 1.8(b), the cone is diverged by the surface, and F is negative.

The *linear magnification* y'/y (Fig. 1.11) due to a spherical refracting surface is given, for a paraxial cone of rays, by:

$$m = \frac{y'}{y} = \frac{x'\theta'}{x\theta}$$

Now, from Snell's law,

$$\frac{\sin \theta}{\sin \theta'} = \frac{\theta}{\theta'} = \frac{n'}{n}$$

Therefore,

$$m = \frac{y'}{y} = \frac{x'\theta'}{x\theta} = \frac{n/x}{n'/x'} \tag{1.7}$$

If the object space medium and the image space medium are the same, $n = n'$ and $m = x'/x$.

It follows from Equation 1.5 that the linear magnification m is numerically equal to the initial vergence V divided by the final vergence V', i.e.

$$m = \left| \frac{V}{V'} \right|$$

for a single surface. Furthermore, as the image produced by one surface forms the object for the next surface, the linear magnification produced by a series of coaxial spherical surfaces is simply the continued product of the V/V' terms for all the surfaces. In general, therefore,

$$m = \left| \prod_{p=1}^{p=N} \frac{V_p}{V'_p} \right| \tag{1.8}$$

where the magnification due to the pth surface is $|V_p/V'_p|$, and the product is taken over N surfaces.

The above discussion regarding the magnification would apply equally well if the surfaces were replaced by optical elements (a lens or a system of lenses); the principal planes of a given element would then replace the single spherical surface.

1.6 The Focal Length and the Power of a Thin Lens

A thin biconvex spherical lens is made of a transparent material of refractive index n_m. An object at a distance x from the first convex face of the lens is in a medium of refractive index n. The radius of curvature of this convex face is r. If the second opposite face were absent and light simply entered the material of refractive index n_m, an 'intermediate image' would be formed at a distance x_i given, from Equation 1.3, by:

$$\frac{n_m}{x_i} + \frac{n}{x} = \frac{n_m - n}{r}$$

However, the second face, which is of radius of curvature r', is present and is concave to the incident light; a final image will, therefore, be found in the medium of refractive index n' [Fig. 1.12(a)] at a distance x' given, from Equation 1.4, by

$$\frac{n'}{x'} - \frac{n_m}{x_i} = \frac{n' - n_m}{r'}$$

the rays now passing from the intermediate image at distance x_i, within a medium of refractive index n_m and emerging into the medium of the image space of refractive index n'. From these two equations it is simple to eliminate n_m/x_i to give:

$$\frac{n'}{x'}+\frac{n}{x} = \frac{n_m-n}{r}+\frac{n'-n_m}{r'} \tag{1.9}$$

If the object and image space media are the same, so that $n' = n$, this equation reduces to:

$$\frac{n}{x'}+\frac{n}{x} = (n-n_m)\left(\frac{1}{r'}-\frac{1}{r}\right)$$

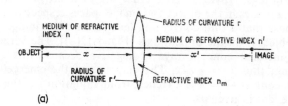

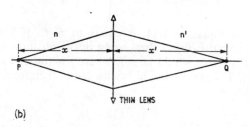

Fig. 1.12. *A thin converging lens: (a) formation of an image; (b) power of the lens*

If x is infinity—i.e. if the object is at an infinite distance from the lens—the image is at the second focal point of the lens; thus $x' = f'$, the second focal length. Substitution of $x = \infty$ and $x' = f'$ into Equation 1.9 gives:

$$\frac{n'}{f'} = \frac{n_m-n}{r}+\frac{n'-n_m}{r'} \tag{1.10}$$

So Equation 1.9 can be written as:

$$\frac{n'}{x'}+\frac{n}{x} = \frac{n'}{f'} \tag{1.11}$$

If the object space and image space media are of the same refractive index n, Equation 1.11 becomes:

$$\frac{1}{x'} + \frac{1}{x} = \frac{1}{f}$$

Equation 1.11 is the conjugate relationship between the image distance x' and the object distance x, the rays of light being paraxial. The fact that the same equation applies to a biconcave lens or a concavo-convex one can be readily verified by similar reasoning. This equation also applies to the general coaxial optical system; such a system may be regarded as replaced by an equivalent thin lens, provided that x and x' are then measured from the principal planes in the object and image spaces respectively.

Suppose that a thin converging lens forms a real image Q of a point object P on the optical axis [Fig. 1.12(b)]. The vergence V of the incident cone of rays is negative because this cone is diverging; thus $V = -n/x$. The vergence of the emergent convergent cone is $V' = n'/x'$. Hence, from Equation 1.6, the power F of the lens is given as

$$F = V' - V = \frac{n'}{x'} - \left(-\frac{n}{x}\right) = \frac{n'}{x'} + \frac{n}{x}$$

which, from Equation 1.11, is equal to n'/f'. If the ray-trace were reversed, the object then being in the medium of refractive index n', the result would be n/f, where f is the first focal length of the lens. The power would be the same; therefore,

$$F = \frac{n'}{f'} = \frac{n}{f}$$

For a thin lens in air, for which $n = n' = 1$, the power is the reciprocal of the focal length. Thus $F = 1/f$ and is positive for a lens that converges an incident beam (for example, a biconvex lens) and negative for a lens that diverges an incident beam (for example, a biconcave lens). If the focal length f is expressed in metres, the power $F = 1/f$ is expressed in *dioptres*.

Note that the use of a sign convention has been avoided in deriving Equations 1.3, 1.4 and 1.11; the problems have been considered solely on geometrical lines. The *optical sign convention* to be used is introduced in conjunction with the concept of vergence. This convention is simply that vergence is positive for a convergent cone of rays, negative for a divergent one; and that if an optical system converges (or renders less divergent) an incident beam, it has a positive power and a positive focal length, whereas if it diverges (or renders less convergent) the incident beam, both the power and the focal length are negative.

1.7 The Power of a System of Coaxial Refracting Elements

In Fig. 1.13, the focal and principal points of a complex system are denoted by F and P respectively in the object space, and F' and P' in the image space. Let F_1 and P_1 denote the first focal and principal points respectively of the first element of the system (which can be, for example, a thin lens). For the whole system, the focal lengths in the object and image spaces are f and f' respectively. Let the first focal length of the first element be f_1, its second focal length be f_1' and, assuming there are N elements, let the second focal length of the final Nth element be f_N'.

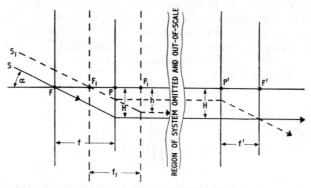

Fig. 1.13. The power of a system of coaxial refracting elements

Consider two parallel incident rays SF and S_1F_1 from a distant point making an angle α with the optical axis. Suppose SF gives rise to a point image in the second focal plane of the whole system at a height H from the optical axis; this will also be the height of the emergent ray at the second principal plane P' of the whole system. The ray S_1F_1 parallel to SF will give rise to a point image in the second focal plane of the first element at a height h above the optical axis, where h is also the height of the emergent ray at the first principal plane of the first element. It follows that:

$H = h \times$ the magnification due to all elements except the first one

From Equation 1.8, it is seen that:

$$H = h \left| \prod_{p=2}^{p=N} \frac{V_p}{V_p'} \right|$$

For paraxial rays, $\tan \alpha = \alpha$; so $H = f\alpha$ and $h = f_1\alpha$. Hence,

$$f = f_1 \left| \prod_{p=2}^{p=N} \frac{V_p}{V_p'} \right| \qquad (1.12)$$

From the definition of the power F of an optical system (Equation 1.6),

$$F = V' - V = \frac{n'}{x'} - \frac{n}{x}$$

where x is the distance of the object from the first principal plane of the whole system, n is the refractive index of the medium of the object space, x' is the conjugate image distance from the second principal plane of the whole system, and n' is the refractive index of the medium of the image space. But $x' = f'$ when $x = \infty$; so for an incident beam from a very distant object, $V = n/x = 0$ and $F = V' = n'/f'$, which is also equal to n/f.

From Equation 1.12,

$$\frac{n}{f} = n \left(f_1 \prod_{p=2}^{p=N} \frac{V_p}{V_p'} \right)^{-1}$$

Therefore,

$$F = F_1 \left(\prod_{p=2}^{p=N} \frac{V_p}{V_p'} \right)^{-1}$$

where $F_1 = n/f_1$ is the power of the first element. If light travelling in the opposite direction through the system (from right to left) were considered, it would be equally true to write:

$$F = F_N \left(\prod_{p=N-1}^{p=1} \frac{V_p}{V_p'} \right)^{-1}$$

Summarising the above results, the power of the general system is given by:

$$F = \frac{n}{f} = \frac{n'}{f'} = F_1 \left(\prod_{p=2}^{p=N} \frac{V_p}{V_p'} \right)^{-1} = F_N \left(\prod_{p=N-1}^{p=1} \frac{V_p}{V_p'} \right)^{-1} \qquad (1.13)$$

Note that if $n = n'$, i.e. if the refractive indices of the object and image space media are the same (as frequently happens because they are both air), $f = f'$, i.e. the two focal lengths are equal.

If the powers of the separate elements of an optical system are known, Equation 1.13 enables the power and focal lengths of the whole system to be found, provided that the terms V_p and V_p' are known for all the elements. To find V_p and V_p', it is necessary to know the relationship between the vergence of the cone of rays leaving one element and the vergence of the same cone arriving at the following element; this problem is considered in the next section.

1.8 The 'Reduced' Vergence of a Cone of Rays Passing between Two Separated Optical Elements

Referring to Fig. 1.14, the relationship is required between the vergence V_2 of a cone of rays arriving at the first principal plane of an optical element 2 and the vergence V_1' of the *same* cone when leaving the second principal plane of the element 1.

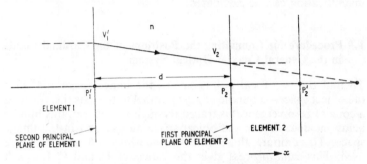

Fig. 1.14. The 'reduced' vergence of a cone of rays passing between two separated optical elements

Let n be the refractive index of the medium between the optical elements 1 and 2. Then,

$$V_1' = \frac{n}{x'} \quad \text{and} \quad V_2 = \frac{n}{x'-d}$$

where x' is the distance along the optical axis from P_1' (the second principal point of element 1) to an axial image point which would be formed by a cone of vergence V_1', and d is the separation along this axis between P_1' and P_2 (the first principal point of element 2). Therefore,

$$V_2 = \frac{n/x'}{1-d/x'} = \frac{V_1'}{1-V_1'(d/n)}$$

which is the relationship required. If this is extended to the general case, the relationship between the vergence V_p of a cone of rays arriving at the first principal plane of the pth optical element and the vergence V_{p-1}' of the cone of rays leaving the second principal plane of the immediately preceding $(p-1)$th element is

$$V_p = \frac{V'_{p-1}}{1 - V'_{p-1}(_{p-1}d_p/_{p-1}n_p)} \tag{1.14}$$

where $_{p-1}d_p$ is the separation along the optical axis between the second principal plane of the $(p-1)$th element and the first principal plane of the pth element, and $_{p-1}n_p$ is the refractive index of the optical medium between these same principal planes.

The use of Equations 1.6 and 1.14 enables any cone of rays to be traced through a system of coaxial optical elements, provided that the power and principal plane positions of each element are known. Furthermore, by means of Equation 1.8, the image magnification can be calculated.

1.9 Procedure for Computing the Positions of the Cardinal Points in the General Coaxial Optical System

To find the two focal points—F' in the image space and F in the object space—a bundle of rays parallel to the axis (i.e. forming a cone of zero vergence) is traced through the system, this bundle being incident first in the object space and second in the image space. To compute these traces, Equations 1.6 and 1.14 are used. This same process gives the values of V_p and V'_p for each optical element p.

To find the principal points, the power of the system is computed by the use of Equation 1.13, and hence the focal lengths f and f' are obtained. The sign of the power indicates the positions of the principal points in relation to the focal points: for a positive or converging optical system, the focal points are outside the principal points, just as they would be for a converging lens; for a negative or diverging optical system, the focal points are between the principal points, as they are for a diverging lens. Once f and f', the positions of the focal points and the sign of the power are known, the positions of the principal points can be found.

When the positions of the focal and principal points have been determined, the nodal points can be found by a simple geometrical method. Let F and P be the first, and F' and P' the second, focal and principal points respectively of an optical system (Fig. 1.15). Suppose the first and second nodal points are known to be at N and N' respectively. Let AB be a ray incident on the optical system towards the first nodal point N, A being in the first focal plane and B in the first principal plane. This ray will emerge from the system as CE, passing through the second nodal point N',

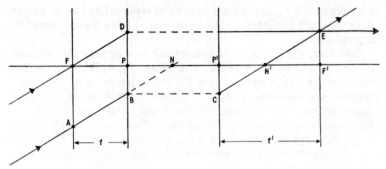

Fig. 1.15. Determination of the nodal points

where C and E are on the second principal and focal planes respectively. From the definition of nodal points, $\angle PNB = \angle EN'F'$. A constructional ray FD incident on the optical system parallel to AB is shown in Fig. 1.15; this ray proceeds to E on emergence. The right-angled triangles FDP and $N'EF'$ are congruent because, since ED is parallel to the optical axis, $\angle DFP = \angle PNB = \angle EN'F'$ and $DP = EF'$. Hence, $N'F' = FP = f$, the first focal length of the system. By a similar argument, $NF = F'P' = f'$, the second focal length of the system.

The results can be summarised as: the distance of the *second* nodal point from the *second* focal point equals the *first* focal length; whilst the distance of the *first* nodal point from the *first* principal point equals the *second* focal length. In each instance, the distances from the focal points to the nodal points are in the same sense as the distances from the principal points to the focal points, as may be seen from Fig. 1.15.

If $f = f'$, as happens when the object and image space media are the same (Section 1.7), the nodal points coincide with the principal points.

1.10 A System Comprising Two Coaxial Optical Elements

A system frequently encountered in practice consists of two optical elements (in the simplest case, two thin lenses) separated by a distance on a common axis. For simplicity, the system is treated as if each element were represented by a single surface. Thus the two optical elements are considered to be at the positions L_1 and L_2 in Fig. 1.16. If these elements were both thin lenses, this simplified method would suffice; if each element were complex (for example, each a thick lens or each two thin lenses in an eyepiece), then L_1, say, would have to be replaced by two principal points and all distances would be measured from those points for a given element.

Consider, then, a system comprising two optical elements of powers F_1 and F_2 separated by a distance $_1d_2$; n_1 is the refractive index of the object space, $_1n_2$ is the refractive index of the space between the elements, and n_3 is the refractive index of the final image space. The positions of the focal points are found, as described in section 1.9, by tracing parallel bundles of rays through the optical system from each axial direction. Starting from the left (the object space), the vergence V_1 of the cone of rays incident

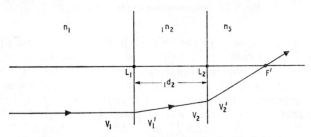

Fig. 1.16. A coaxial system of two separated optical elements

on element L_1 is zero because the 'cone' is a bundle of rays parallel to the optical axis. The vergence V_1' on leaving L_1 is given by:

$$V_1' = V_1 + F_1 = F_1 \tag{1.15}$$

The vergence V_2 of the cone arriving at L_2 is given, from Equation 1.14, by:

$$V_2 = \frac{V_1'}{1 - V_1'(_1d_2/_1n_2)} = \frac{F_1}{1 - F_1(_1d_2/_1n_2)} \tag{1.16}$$

The vergence V_2' of the cone leaving L_2 is given by:

$$V_2' = V_2 + F_2 = \frac{F_1}{1 - F_1(_1d_2/_1n_2)} + F_2 = \frac{F_1 + F_2 - F_1F_2(_1d_2/_1n_2)}{1 - F_1(_1d_2/_1n_2)} \tag{1.17}$$

A similar computation can be undertaken for a parallel bundle of rays from the right, i.e. from the image space.

It follows that the distance v' from L_2 to the second focal point is given by

$$\frac{n_3}{v'} = \frac{F_1 + F_2 - F_1F_2(_1d_2/_1n_2)}{1 - F_1(_1d_2/_1n_2)} \tag{1.18}$$

because $V_2' = n_3/v'$ (see Equation 1.5); also, the distance v from L_1 to the first focal point is found by computing the ray trace in the opposite direction. Thus,

$$\frac{n_1}{v} = \frac{F_1 + F_2 - F_1 F_2({}_1 d_2/{}_1 n_2)}{1 - F_2({}_1 d_2/{}_1 n_2)} \tag{1.19}$$

The sign of each final vergence will determine whether the emergent beam converges to a real focal point or diverges from a virtual one.

When the focal points have been found from Equations 1.18 and 1.19, the power F of the system can be calculated. From Equation 1.13,

$$F = \frac{n_1}{f} = \frac{n_3}{f'} = F_1 \left(\frac{V_2}{V_2'}\right)^{-1}$$

Therefore, from Equations 1.16 and 1.17,

$$F = F_1 \left[\frac{F_1}{1 - F_1({}_1 d_2/{}_1 n_2)}\right]^{-1} \left[\frac{1 - F_1({}_1 d_2/{}_1 n_2)}{F_1 + F_2 - F_1 F_2({}_1 d_2/{}_1 n_2)}\right]^{-1}$$

Therefore,

$$F = F_1 + F_2 - \frac{{}_1 d_2}{{}_1 n_2} F_1 F_2 \tag{1.20}$$

It follows that, since $f = n_1/F$ and $f' = n_3/F$, both f and f' can be calculated. The positions of the principal points and the nodal points can, therefore, be found by the methods described in Section 1.9 or by the geometrical method of Section 1.11.

If F is positive, the system is converging; if F is negative, the system is diverging.

1.11 A Geometrical Method for Finding the Positions of the Principal Points in the Two-element Coaxial System

After the focal points of the two-element system have been located, the principal points may be alternatively found by the following geometrical method. Again, L_1 and L_2 represent the two optical elements in Fig. 1.17. If the incident ray parallel to the axis is produced until it meets the emergent ray BF' produced backwards, a point C in the second principal plane will be located. From the vergence theorem, the distance f_1' from the first element to its second focal point F_1' can be found, and also the distance v' of the second focal point F' of the system from the second

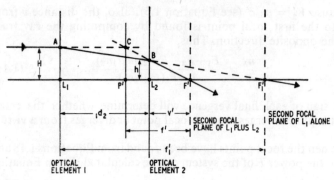

Fig. 1.17. Geometrical method of locating the principal points in the general two-element system

element L_2. As the triangles $AF_1'L_1$ and $BF_1'L_2$ are similar,

$$\frac{H}{h} = \frac{f_1'}{f_1' - {}_1d_2}$$

where $H = AL_1$ and $h = BL_2$. Furthermore, triangles $CF'P'$ and $BF'L_2$ are also similar and $CP' = H$; therefore,

$$\frac{H}{h} = \frac{f'}{v'}$$

From these equations it follows that:

$$f' = \frac{v'f_1'}{f_1' - {}_1d_2} \tag{1.21}$$

Similarly, by tracing a ray parallel to the axis from the opposite direction, the first focal length f may be found, and the two principal points are thus located.

1.12 Location of the Cardinal Points by Experiment

To locate completely the three pairs of cardinal points of an optical system, it is sufficient to determine the positions of the focal points and either the principal points or the nodal points. Thus, a knowledge of the positions of the focal and principal points leads to easy calculation of the two focal lengths and so also of the positions of the nodal points (Section 1.9). Similarly, if the focal and nodal points are located, the positions of the principal points may be calculated.

1.12.1 LOCATION OF THE FOCAL POINTS

To locate the focal points F and F' of a converging optical system, a collimated beam of light parallel to the optical axis is directed through the system from one side, and the focus is recorded on a screen on the other side. If the incident light is in the object space, the screen locates the second focal point F'; if the light direction is reversed (i.e. if it is incident in the image space), the first focal point F is located. This is, of course, the same procedure as for a thin converging lens. Alternatively, and again as for a thin lens, an object in the form of an illuminated cross-wire in an aperture in a white screen can be used. On the other side of the lens system is then situated a plane mirror normal to the optical axis. The object screen is moved up and down the optical axis until the image of the cross-wire is focused to coincide with the object aperture. The object screen is then in one of the focal planes of the lens system.

1.12.2 LOCATION OF THE PRINCIPAL POINTS

A useful procedure for finding the principal points of a converging system is *Newton's method*. An illuminated cross-wire in an aperture is used as the object; a real image of this object is focused on a screen on the other side of the optical system. A series of

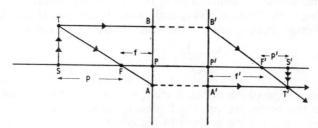

Fig. 1.18. Newton's method

conjugate object distances p and image distances p' measured from the respective focal points F and F' are found. Referring to Fig. 1.18, in which ST is an object giving rise to an image $S'T'$ and the familiar ray-trace has been drawn, it is clear that triangles TSF and APF are similar. Therefore,

$$\frac{p}{f} = \frac{ST}{PA}$$

Again, since triangles $T'S'F'$ and $B'P'F'$ are similar,

$$\frac{p'}{f'} = \frac{S'T'}{B'P'}$$

Further, as $PA = P'A' = S'T'$ and $ST = BP = B'P'$,

$$\frac{p}{f} = \frac{f'}{p'}$$

or

$$pp' = ff' \tag{1.22}$$

In the usual situation where the object and image media are the same (for example, air) $f = f'$, so $pp' = f^2$. Hence, a linear plot of p against $1/p'$ for a series of positions of the object and corresponding image locations will give f^2. Once f is known, the principal points P and P' can be obtained; and, since the object and image media are the same, the principal points and the nodal points will coincide (Section 1.9).

1.12.3 MAGNIFICATION METHODS

The most convenient magnification methods employ an illuminated transparent scale as the object and a similar scale for the image screen. Alternatively, a travelling microscope may be used for measuring the dimensions of the object and image. If a converging optical system giving real images is used and if the object and image space media are the same, Equation 1.11 gives

$$\frac{1}{x'} + \frac{1}{x} = \frac{1}{f}$$

where x is the object distance and x' the image distance measured from the respective principal planes. Also, where $n = n'$, the linear magnification is given by Equation 1.7 as:

$$m = \frac{x'}{x}$$

From these two equations it follows that:

$$m + 1 = \frac{x'}{f} \tag{1.23}$$

Now $x' = l + z$, where l is the distance of the image from any suitable fixed point on the optical axis (best measured on an optical bench), and z is the distance from this fixed point to the second principal point P'. So Equation 1.23 becomes:

$$m+1 = \frac{l+z}{f}$$

As z is clearly a constant for a fixed position of the optical system, a straight line will be obtained if m is plotted against l, the slope of the line being $1/f$.

Alternatively, referring to Fig. 1.19 and using similar triangles:

$$\frac{h'}{h} = \frac{x'-f}{f} = \frac{f}{x-f} \tag{1.24}$$

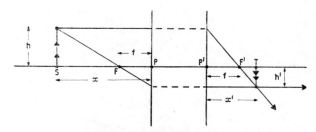

Fig. 1.19. The magnification method of determining the cardinal points of an optical system

If x is changed by a distance x_1, x' will change by a distance x_2 and the new image formed will be of height h''. Equation 1.24 will then become

$$\frac{h''}{h} = \frac{x'+x_2-f}{f} = \frac{f}{x+x_1-f} \tag{1.25}$$

where x_1 and x_2 may be positive or negative. As x_1 and x_2 are changes of distance along an optical bench in the experiment, they may be measured accurately; therefore, if h, h' and h'' are also determined, f can be found from Equations 1.24 and 1.25.

1.12.4 SEARLE'S GONIOMETER

A simple apparatus for finding the focal length of a converging lens system in air is Searle's goniometer [Fig. 1.20(a)]. This consists of a T-shaped fixed base, usually of wood, the member AB being marked with a linear graduated scale which has a centre zero. A beam CD, which can be rotated about a pivot, rests on this fixed base. On CD, and above the pivot, is mounted an

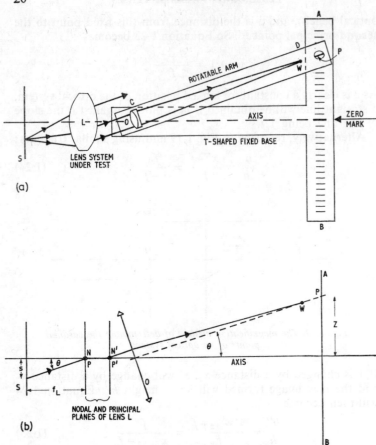

Fig. 1.20. Searle's goniometer

achromatic objective O; and at a distance along the beam axis, equal to the focal length of this objective, is a fine vertical wire W in a supporting open frame. The beam CD also contains an aperture with a pointer P arranged above the scale in AB. When P is above the zero mark of the scale in AB, the axis of the rotatable beam arm CD is aligned with the axis of symmetry of the fixed T-shaped support. Initially, the instrument is best set with P over this zero mark. The converging lens system L under test is then placed in a support between a second scale S and the objective O, as shown in Fig. 1.20(a). S is conveniently an illuminated glass scale.

The scale S is moved along the optical axis of L (the lens system under investigation) until it is in the focal plane of L. This position of S is located when no parallax can be seen between its image (as formed by both L and O) and the vertical wire W. Clearly, the light leaving L is then forming a parallel beam which is incident on O and is brought to a focus at W. The scale reading on S is taken in this position, and the beam CD is then rotated until a second division of the scale S forms an image at W. The difference is noted between the two readings on the scale S, corresponding to an object of linear dimension s. The angle θ through which CD has been rotated is known from the separation z between the initial and final positions of P on the scale in AB, in conjunction with the fixed distance from the pivot at O to the pointer P.

From the ray diagram of Fig. 1.20(b) it can be seen that

$$\tan \theta = \frac{s}{f_L}$$

where s is the known difference between the readings on the scale S, and f_L is the focal length of the lens L—which is thus determined. In practice, the angle θ is kept small as otherwise distorted images may be produced; so $\tan \theta = \theta$ and $f_L = s/\theta$.

1.12.5 THE NODAL SLIDE

The nodal slide (Fig. 1.21) enables the nodal points and focal lengths of a converging lens system to be found. A students' nodal slide is shown in Fig. 1.21(a). The lens system is mounted in supports on a horizontal platform which can be rotated about a vertical axis through its centre (the position of which is marked). On this platform, and parallel to the optical axis of the lens system, is a scale which can be used, for example, to measure the separation between two thin lenses forming the optical system. This scale can also be moved, together with the supports to the lens system, perpendicularly to its length across the platform and clamped in any position.

In order to find the nodal points of the lens system, its supports and the scale are first moved laterally across the horizontal platform (i.e. perpendicularly to the optical axis of the lens system) until this optical axis is judged by eye to intersect the vertical axis of rotation of the platform, and to be in line with a collimated beam of light from a projector. To check the correct setting of both the axis of the collimated beam and the optical axis relative to the axis of rotation, the position on a screen is noted

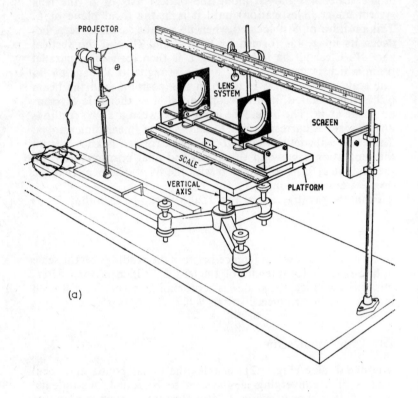

(a)

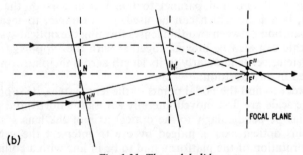

(b)

Fig. 1.21. The nodal slide

of the image formed by the lens system of cross-wires in the projector. The lens system under investigation is then rotated through 180° about the vertical axis, and the new position of the image noted. The lens system and the collimated light beam are next adjusted laterally until the image is midway between the two previous positions.

Once the correct setting has been obtained, the nodal and focal points can be found in the following way. The lens system is moved on the platform along its optical axis until the position is found at which a small rotation of this system about the vertical axis causes a minimum lateral displacement of the image on the screen. The vertical axis through the horizontal platform now intersects a nodal point of the system, say N', and the incident parallel beam is focused at the corresponding focal point F'. The ray-trace concerned here is shown in Fig. 1.21(b): the small rotation about the second nodal point N' causes the first nodal point N to move to N''; by definition, the ray from the incident collimated beam through N'' must emerge undeviated along $N'F'$ parallel to its initial direction. At the screen within the focal plane, the focus will now appear at F'' very close to F' and, indeed, F'' and F' will coincide in effect if the rotation is small. Thus, the incident collimated beam produces an image which is undisplaced by small rotation.

To find the first nodal and focal points, N and F respectively, the whole system on the platform is rotated through 180° and the procedure previously used to find N' and F' is followed.

A separate horizontal scale is used to find the focal length $N'F'$ or NF, which in both cases is the separation between the vertical axis of rotation of the platform and the image on the screen.

Instead of using a projector as a source of collimated light for the nodal slide, an auto-collimating method may be employed; in this, an illuminated aperture in a screen is used as the object, and a plane mirror is placed perpendicular to the optical axis on the other side of the lens system. The lens system supports and the scale are then moved laterally across the platform until the image coincides with the object; and, on small rotations about the vertical axis as above, this image is not displaced.

1.13 Stops in an Optical System

In an optical system, the term 'stop' is generally used to denote a means of restricting the angle of convergence or divergence of a cone of light rays passing through the system. As the great majority of optical systems consist of coaxial spherical lenses or mirrors, stops are usually in the form of a circular aperture in a thin plate placed with its centre at a carefully chosen point on the optical axis. However, *any* element of a system may restrict the cone of light and is therefore a stop. The function of a stop in restricting the angles about the optical axis of the cones of light which pass through the system is primarily to ensure that optical aberrations are minimised. Such stops inevitably also reduce the amount of light passing through the system from object to image, and they also reduce the resolving power of the system. A common example of a variable stop (i.e. one of which the aperture can be altered at will to vary the amount of light passing through the system) is the iris diaphragm used in a

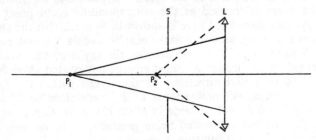

Fig. 1.22. The aperture stop

camera lens. Even if no actual stop in the form of an aperture in a diaphragm is inserted within the train of lenses, a stop will be present and decided by the apertures of the lenses themselves. Again, a stop may be in the form of an annular ring within an opaque plate; moreover, circular symmetry is not inevitable in that a rectangular slot or any other geometry of aperture may be concerned.

The stop which limits the angle of an axial cone of rays passing through an optical system is more specifically called an *aperture stop;* it is illustrated in the simple lens system of Fig. 1.22. Here, the stop S limits the vergence of the axial cone from the point object P_1. Note, however, that the lens L itself would form the aperture stop of the system if the object were at P_2. Indeed, the angular aperture of the stop manifestly depends on the position of the object on the optical axis.

The iris diaphragm usually employed between the components of a camera objective lens has a *relative aperture* denoted by an f-number N. This number is given by dividing the focal length f of the lens by the diameter d of the exit pupil, which is defined below. For example, N may be 2, 5·6, 8 and 11 corresponding to diameters d of the exit pupil due to the aperture in the diaphragm being respectively 1/2, 1/5·6, 1/8 and 1/11 times the focal length of the lens.

To determine the amount of light which enters an optical system, it is convenient to define the *entrance pupil* of a coaxial system of optical elements (for example, a microscope or telescope). The entrance pupil is the image of the aperture stop in all the optical elements which precede it, i.e. are between it and the object. This pupil determines the angle of the cone of light rays which enters the system and so decides its 'light gathering' power. For example, consider an optical system consisting of two coaxial separated thin lenses L and L' (Fig. 1.23). The aperture stop is at S, between L and L', and the point object observed through the system is at O. The image of S in the optical elements preceding it (in this example, lens L alone) is S', which is the entrance pupil of the system and decides how much light from O enters the system. S' is indicated in the diagram between L and L', but this is only for convenience: depending on the lenses and their separation, S' may be anywhere between O and the final image I.

The entrance pupil is alternatively and better defined as that pupil or aperture which subtends the smallest angle at the object point. In practice, the entrance pupil of a given lens system (which may or may not include actual diaphragm stops) is required so that the amount of light entering the system from an

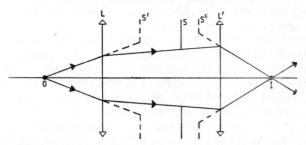

Fig. 1.23. The entrance and exit pupils of an optical system

object point on the optical axis can be estimated. There are a number of possible aperture stops in the optical system; they may be lenses, mirrors, actual diaphragm stops or, if the design is bad, simply some ring on the surface of the cylindrical barrel in which the optical elements are mounted. All possible aperture stops are explored, and their corresponding entrance pupils are found by determining their images in those optical elements which precede them. The entrance pupil which subtends the smallest angle at the object point is then found; this is clearly the one which affects how much light enters the system. This particular pupil is the entrance pupil proper of the system.

The *exit pupil* of an optical system is concerned with how much light emerges from the system. It is defined as the image of the aperture stop formed by all the optical elements which follow it, and determines the maximum possible angle of any cone of rays which leaves the system to form the image. The entrance and exit pupils together affect the brightness of this final image.

For evaluating the size of the exit pupil of a specific system, it is better defined as that pupil in the system which subtends the least angle at the axial image point. In Fig. 1.23, S'' is the image of S formed by L' and is the exit pupil of the system.

In elementary optics, the usual practice in studying image formation is to draw a ray-trace diagram using certain single selected rays from the object. Though the image position and size may be located in such a way, information about the amount of light which forms the image is not given. To consider this light flux, which decides the image brightness, *cones* of rays passing through the system must be evaluated in relation to entrance and exit pupils.

In discussing the passage of such cones of rays, the concept of the *chief ray* is useful. This is defined for any cone as that ray which passes from a point on the object to a point on the image through the centre of the entrance and exit pupils, and so through the centre of the aperture stop.

Apart from restriction on image brightness due to stops, there will usually be a stop in the optical system which limits the field of view: this is called the *field stop*. This is important in forming an image of an extended object; the field stop—which may not be an actual stop but some obstruction in the mount, or the surrounds of the optical elements themselves or, for example, the extent of the photographic plate in a camera—may allow the central regions of the object to be imaged with adequate brightness but not the outer regions of the object.

Consider again two coaxial separated lenses L_1 and L_2 [Fig. 1.24(a)] through which an extended object is viewed. Suppose lens L_1 is 'stopped down' (for example, with an iris diaphragm) so that its aperture stop is between the short horizontal lines S and S'. This aperture SS' is also the entrance pupil of the system because there is no lens preceding it. *It is the second lens L_2, of diameter Q_1Q_2, which forms the field stop of this system:* the extremities of the field of view are decided by the rays OQ_1 and OQ_2. If OQ_1 and OQ_2 are both prolonged, they decide the chief rays P_1OQ_1 and P_2OQ_2. Thus, the whole of the object is not imaged satisfactorily because it extends outside P_1 and P_2.

In fact, the field of view *will* extend outside P_1 and P_2, but cones from such outside points will inevitably give rise to fainter images than those from within the field of view proper.

Cones of rays from P_1 and P_2 at the extremities of the field of view proper will be approximately half-transmitted through lens

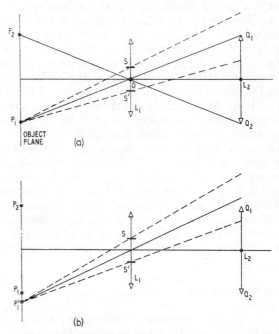

Fig. 1.24. *The concept of the field stop*

L_2. Thus, the dotted lines of Fig. 1.24(a) represent a cone symmetrical about the chief ray P_1Q_1; it is clear that it is approximately half-transmitted through L_2. On the other hand, a symmetrical cone about a ray from an extreme point P_1' outside P_1 in Fig. 1.24(b) will obviously be less than half-transmitted by lens L_2, and so will give rise to a fainter image. It is, therefore, convenient to fix the limit of the field of view as corresponding to the chief rays of cones that are half-transmitted. In a camera, for example, it is important that the field of view be limited by the plate or film to ensure uniformity of brightness over this plate of an image from a uniformly bright extended object.

The semi-angle subtended by the field of view at the entrance pupil [i.e. $\angle Q_1 O L_2$ in Fig. 1.24(a)] is called the *semi-angular field of view;* it is consequently half the angle subtended at the entrance pupil by the field stop.

The entrance and exit pupils of an optical system may be regarded as convenient means of indicating the maximum angles of cones of rays entering and leaving the optical system. However, in view of the presence of the field stop, it is necessary to define a further pair of pupils which show the effect of this stop on these cones. These pupils are called the *entrance* and *exit windows* of the system. They are defined, respectively, as the image of the field stop in all the optical elements before it, and the image of the field stop in all the optical elements after it. The entrance window must, therefore, subtend twice the semi-angular field of view at the centre of the entrance pupil and, if the positions and sizes of both the entrance and exit windows are known, it is possible to determine completely the effect of the stops in the system on cones of rays passing through it.

In practice, the stop in the system which actually forms the field stop may not be known. To find it, all possible locations of the entrance window, corresponding to various possible positions of the field stop, are explored and the angles that these entrance windows subtend at the entrance pupil are found. The least angle so determined gives the entrance window corresponding to the field stop proper of the optical system. The operations of finding the field stop and the aperture stop may thus be combined.

1.14 Telecentric Systems

A telecentric optical system is one in which the centre of the aperture stop is at a principal focus of the system. If it is at the second principal focus, in the image space, the entrance pupil of the system must be at infinity on the object side since, by definition, this pupil is the image of the aperture stop in all the optical elements which precede it. The incident chief ray is thus parallel to the optical axis, and the system is said to be telecentric on the object side, as in the example of the single convergent objective lens shown in Fig. 1.25(a). In this diagram, the aperture stop centre is at the second principal focus of the objective lens; this is also the exit pupil of this lens because no optical elements follow it on the image side. Consider cones of rays from points P_1 and P_2 in the plane of the extended object. These give rise to image plane points Q_1 and Q_2 respectively (cones of rays from P_1 and P_2 about the chief rays are shown by the shaded areas in the diagram). If the object is not exactly positioned to give a focused image in the image plane (which is predetermined, for example, by cross-wires in an instrument), the apparent positions of the points P_1 and P_2 in the object are unaltered because the chief

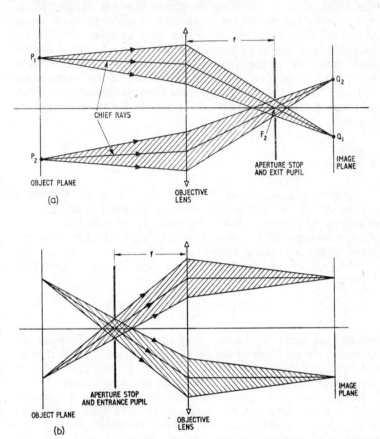

Fig. 1.25. Single-lens telecentric systems: (a) telecentric on the object side; (b) telecentric on the image side

rays of the cones from them are parallel to the optical axis. A system telecentric on the object side is therefore valuable in a travelling microscope or telescope (cathetometer). If the distance P_1 to P_2 is to be measured by suitable traverse of the microscope, the measurement obtained is unaffected if P_1 and P_2 are not precisely focused.

An optical system may also be telecentric on the image side, as in the single objective lens system of Fig. 1.25(b). Here, the aperture stop, now the same as the entrance pupil, has its centre at the first principal focus on the object side. The chief rays of the cones of rays shown on the image side are parallel to the optical axis, and the system is said to be telecentric on the image side. This kind of system, but with a more complex train of co-axial lenses, is useful in a microscope furnished with a divided scale (graticule) in the eyepiece against which the separation of two points in an object are to be measured; if the image is not precisely focused at the graticule, the observations are not affected because the chief rays of the cones to the point images are parallel.

Examples

In the worked examples which follow, the concept of vergence of a cone of rays is used, as explained in Section 1.5, and the terminology of Section 1.8 is employed. To recapitulate briefly: V_p, the vergence of a cone of rays arriving at the first principal plane of the pth optical element, is related to the vergence V'_{p-1} of the cone of rays leaving the preceding $(p-1)$th element by Equation 1.14:

$$V_p = \frac{V'_{p-1}}{1 - V'_{p-1}(_{p-1}d_p/_{p-1}n_p)}$$

where $_{p-1}d_p$ is the separation along the optical axis between the second principal plane of the $(p-1)$th element and the first principal plane of the pth element, and $_{p-1}n_p$ is the refractive index of the optical medium between these same principal planes.

Often, only two optical elements separated by a distance $_{p-1}d_p = d$ are involved, the medium between them being air; thus, $_{p-1}n_p = 1$, and the incident vergence on element 2 is $V_p = V_2$, where the emergent vergence from element 1 is $V'_{p-1} = V'_1$. Equation 1.14 then simplifies to:

$$V_2 = \frac{V'_1}{1 - V'_1 d} \qquad (1.26)$$

Note. In these examples and the exercises at the end of each chapter, the abbreviations in brackets at the end of a question denote the university or institution concerned. These abbreviations are as follows.

L.G. B.Sc. General Degree of the University of London

L. P.   B.Sc. (Special) Physics Degree of the University of London

L. Anc. Ancillary examination in Physics to the B.Sc. (Special) Chemistry Degree of the University of London

G. I. P. Graduateship Examination of the Institute of Physics

Aston. B.Sc. (Special) Physics Degree of the University of Aston in Birmingham

M. P.   B.Sc. Honours Degree in Physics of the University of Manchester

> 1. *A coaxial lens system in air consists of a thin converging lens of focal length 15 cm followed at a separation of 5 cm by a converging lens of focal length 10 cm. Find the cardinal points of this system.*

In Fig. 1.26(a), AB is one ray of a bundle of rays parallel to the axis and incident from the left upon the first lens L_1. As L_1 is a thin lens, its principal planes coincide through its optical centre. Then, let V_1 be the vergence incident at the first lens L_1, V_1' the vergence of the cone of rays emergent from L_1, V_2 the vergence incident at the second lens L_2, and V_2' the vergence on emergence from L_2. (This notation, following the general notation of Equations 1.14 and 1.26 will be used in all appropriate optical calculations in this book.)

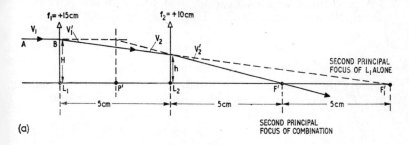

(a)

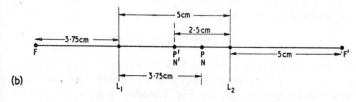

(b)

Fig. 1.26. Example 1

Clearly, $V_1 = 0$ and $V_1' = 1/15$. Since $d = 5$ cm, V_2 is given by Equation 1.26 as:

$$V_2 = \frac{1/15}{1 - (1/15)5} = \frac{1}{10}$$

Alternatively, this may be seen directly from Fig. 1.26(a) because V_2 refers to lens L_2. Finally, $V_2' = 1/v'$, where v' is the separation between L_2 and the second principal focus F' of the system. At lens L_2, Equation 1.6 gives

$$V_2' - V_2 = F_2 = \frac{1}{f_2}$$

where F_2 is the power of lens L_2 of focal length $f_2 = 10$ cm. Therefore,

$$\frac{1}{v'} - \frac{1}{10} = \frac{1}{10}$$

and thus $v' = 5$ cm; i.e. the second principal focus F' of the combination is 5 cm beyond L_2. In a similar manner, a ray parallel to the axis from the right-hand side incident upon lens L_2 may be traced to give $v = 3\frac{3}{4}$ cm; hence, the position of the first principal focus of the combination is $3\frac{3}{4}$ cm in front of L_1.

A geometrical approach can be used to locate the principal points. With reference to Fig. 1.26(a) and by similar triangles,

$$\frac{H}{h} = \frac{15}{10} = \frac{f'}{5}$$

where f' is the second focal length of the combination (equal to $P'F'$, where P' is the second principal point). Therefore,

$$f' = \frac{75}{10} = 7 \cdot 5 \text{ cm}$$

and $P'L_2 = 2 \cdot 5$ cm; i.e. the second principal point is $2 \cdot 5$ cm in front (to the left-hand side) of the lens L_2.

As the system is in the same medium (air) throughout, $f = f'$, where f is the first focal length. Therefore,

$$L_1P = 7 \cdot 5 - 3 \cdot 75 = 3 \cdot 75 \text{ cm}$$

i.e. P, the first principal point, is $3 \cdot 75$ cm beyond (or to the right-hand side of) L_1. The nodal points will coincide with the principal points in this example. The cardinal points are consequently as shown in Fig. 1.26(b).

2. *Define the cardinal points of a lens system.*
   Light from a distant object is incident on a system comprising a thin converging lens of focal length 12 cm placed 6 cm in front of a thin diverging lens of focal length 24 cm. Explain with the aid of a diagram why the position of the image is unaffected if the lens system is rotated about an axis perpendicular to its principal axis and 2 cm in front of the converging lens.
   At what distance from the converging lens do conjugate object and image planes coincide?
   (L.P.)

The definitions of the cardinal points of a lens system are given in Section 1.2.

The point about which the slight rotation occurs must be the second nodal point of the system; the paths of the rays through the optical system would then be modified as shown in Fig. 1.27(a). An incident ray AB parallel to the axis passes, in effect, to the first nodal plane through N after rotation and emerges undeviated through N', the second nodal point; thus, it still meets the focal plane through the second principal focus F'. Consequently, all incident parallel rays meet at F' after refraction, despite the small rotation of the system about N'.

To show that this second nodal point is 2 cm in front of the converging lens and to solve the last part of the question, the cardinal points of the system must be found. The trace of a parallel pencil of rays through the system from one side is shown in Fig. 1.27(b).

With the vergences as denoted in Fig. 1.27(b), $V_1 = 0$, $V_1' = 1/12$, and, from Equation 1.26:

$$V_2 = \frac{V_1'}{1 - 6V_1'}$$

Therefore,

$$V_2 = \frac{1/12}{1 - 6(1/12)} = \frac{1}{6}$$

Also, $V_2' = 1/v'$, where v' is the distance beyond the second lens L_2 to the second principal focus of the combination at F'. But, from Equation 1.6 applied to lens L_2,

$$V_2' - V_2 = \frac{1}{f_2}$$

where $f_2 = -24$ cm, the focal length of the diverging lens L_2. Therefore,

$$\frac{1}{v'} = \frac{1}{6} - \frac{1}{24} = \frac{1}{8}$$

Hence, v', the distance beyond the second lens to the second
principal focus F', is 8 cm.

Similarly, if a pencil of rays is traced in the opposite direction, it is found that $v = 20$ cm, where v is the distance that the
first principal focus F is in front of the first lens L_1.

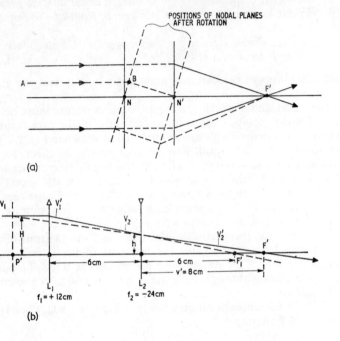

(a)

(b)

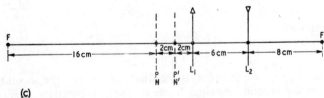

(c)

Fig. 1.27. Example 2

By similar triangles, referring to Fig. 1.27(b),

$$\frac{H}{h} = \frac{12}{6} = \frac{f'}{8}$$

where $f' = P'F'$, the second focal length of the combination.
So $f' = 16$ cm which, since the object and image space media

are the same, is also equal to f. Hence, the cardinal points are as shown in Fig. 1.27(c), and the second principal and nodal points—which coincide—are seen to be 2 cm in front of the converging lens L_1.

To find the position at which conjugate object and image planes coincide, assume this to be at a distance X in front of the first principal point P. An axial object point at this position will create a divergent cone of rays incident on the first principal plane: the vergence of this cone will be $-1/X$, the negative sign being introduced for divergence. The conjugate image point will be in the same position, and so at a distance $X+2$ from the second principal plane P'. The rays from P' travelling towards the right-hand side will clearly diverge, coming from the virtual image on the left. The vergence of this diverging cone relative to the principal plane through P' will therefore be $-1/(X+2)$. The emergent vergence minus the incident vergence is equal to the power of the lens combination which is the reciprocal of its focal length, i.e. $1/16$. Therefore,

$$\frac{-1}{X+2} - \left(\frac{-1}{X}\right) = \frac{1}{16}$$

Thus,

$$\frac{X-(X+2)}{X(X+2)} = -\frac{1}{16}$$

or

$$X^2 + 2X - 32 = 0$$

Therefore, $X = 4{\cdot}744$ cm or $-6{\cdot}744$ cm. The corresponding distances from the converging lens at which conjugate object and image planes coincide are $8{\cdot}744$ cm in front of this lens (to the left-hand side in the diagram) and $2{\cdot}744$ cm behind it. The latter position is inadmissible as it is between the two lenses.

3. *Three coaxial converging lenses L_1, L_2 and L_3 in air, having powers F_1, F_2 and F_3 and corresponding focal lengths f_1, f_2 and f_3, are arranged coaxially as shown in Fig. 1.28. The distance D between the outer components is fixed, but the separation x between the lenses L_1 and L_2 may be varied to alter the total power F of the system. Determine F as a function of x and find the value of x for which F is a maximum.*

Consider a parallel pencil of rays incident from the left-hand side of the diagram. The vergence $V_1 = 0$, and the vergence on leaving L_1 will be given by $V_1' = F_1$. From Equation 1.26, it is seen that the incident vergence V_2 at L_2 is given by

$$V_2 = \frac{V_1'}{1 - V_1' x} = \frac{F_1}{1 - x F_1}$$

the emergent vergence V_2' from L_2 is

$$V_2' = \frac{F_1}{1 - x F_1} + F_2 = \frac{F_1 + F_2 - x F_1 F_2}{1 - x F_1}$$

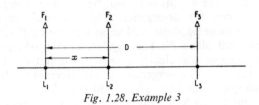

Fig. 1.28. Example 3

and the incident vergence V_3 at L_3 is given by

$$
\begin{aligned}
V_3 &= \frac{V_2'}{1 - V_2'(D - x)} \\
&= \left(\frac{F_1 + F_2 - x F_1 F_2}{1 - x F_1} \right) \left[1 - (D - x) \left(\frac{F_1 + F_2 - x F_1 F_2}{1 - x F_1} \right) \right]^{-1} \\
&= \frac{F_1 + F_2 - x F_1 F_2}{1 - x F_1 - (D - x)(F_1 + F_2 - x F_1 F_2)}
\end{aligned}
$$

Also,

$$
\begin{aligned}
V_3' &= \frac{F_1 + F_2 - x F_1 F_2}{1 - x F_1 - (D - x)(F_1 + F_2 - x F_1 F_2)} + F_3 \\
&= \frac{F_1 + F_2 - x F_1 F_2 + F_3 - x F_1 F_3 - F_3(D - x)(F_1 + F_2 - x F_1 F_2)}{1 - x F_1 - (D - x)(F_1 + F_2 - x F_1 F_2)}
\end{aligned}
$$

Now, from Equation 1.13,

$$F = F_1 \left(\prod_{p=2}^{p=N} \frac{V_p}{V_p'} \right)^{-1}$$

In this example, $N = 3$; therefore,

$$
\begin{aligned}
\prod_{p=2}^{p=3} \frac{V_p}{V_p'} &= \left(\frac{V_2}{V_2'} \right) \left(\frac{V_3}{V_3'} \right) \\
&= \left(\frac{F_1}{1 - x F_1} \right) \times \left(\frac{1 - x F_1}{F_1 + F_2 - x F_1 F_2} \right) \times
\end{aligned}
$$

$$\left[\frac{F_1+F_2-xF_1F_2}{1-xF_1-(D-x)(F_1+F_2-xF_1F_2)}\right]\times$$

$$\left[\frac{1-xF_1-(D-x)(F_1+F_2-xF_1F_2)}{F_1+F_2+F_3-xF_1F_2-xF_1F_3-F_3(D-x)(F_1+F_2-xF_1F_2)}\right]$$

$$=\frac{F_1}{F_1+F_2+F_3-xF_1F_2-xF_1F_3-F_3(D-x)(F_1+F_2-xF_1F_2)}$$

Therefore,

$$F = F_1+F_2+F_3-xF_1F_2-xF_1F_3-F_3(D-x)(F_1+F_2-xF_1F_2)$$
$$= F_1+F_2+F_3-F_1F_3D-F_2F_3D-x(F_1F_2-F_2F_3-DF_1F_2F_3)$$
$$-x^2F_1F_2F_3$$

For maximum power,

$$\frac{\mathrm{d}F}{\mathrm{d}x}=0$$

i.e.

$$x = \frac{-F_1F_2+F_2F_3+DF_1F_2F_3}{2F_1F_2F_3}$$

$$= \frac{1}{2}\left(D-\frac{1}{F_3}+\frac{1}{F_1}\right)$$

Therefore, for maximum power,

$$x = \frac{1}{2}(D-f_3+f_1)$$

4. *A converging meniscus lens 3 cm thick has its concave surface in contact with water of refractive index 4/3 and its convex surface exposed to air. If the material of the lens has a refractive index 3/2 and the radii of curvature of the surfaces are 10 cm and 15 cm respectively, determine the positions of the cardinal points of the system. Illustrate your answer by a diagram drawn roughly to scale and indicate briefly how you would determine the cardinal points experimentally.* (L. P.)

A pencil of parallel rays is traced through the system from the left-hand side in air [Fig. 1.29(a)]. If the usual vergence notation is employed, $V_1 = 0$.

The power of the first convex surface is given by Equation 1.4 to be

$$\frac{n-1}{r} = \frac{3/2-1}{10} = \frac{1}{20}$$

which is positive because the incident beam is converged by this surface. This power is equal to the difference between the emergent vergence V_1' and the incident vergence V_1; therefore,

$$V_1' - V_1 = \frac{1}{20}$$

Therefore $V_1' = 1/20$. Applying Equation 1.14 for $V_{p-1}' = V_1' = 1/20$, $_{p-1}d_p = 3$ cm and $_{p-1}n_p = 3/2$, it follows that:

$$V_2 = \frac{1/20}{1-(3/20)/(3/2)} = \frac{1}{18}$$

Again, the power of the second surface is seen from Equation 1.4 to be

$$\frac{4/3-3/2}{15} = -\frac{1}{90}$$

which is negative as the incident beam is diverged by this surface.

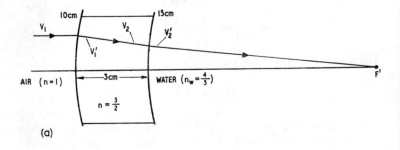

(a)

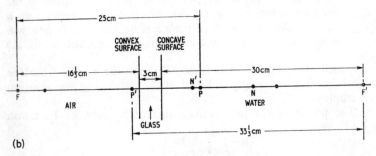

(b)

Fig. 1.29. Example 4

Therefore, V_2', the emergent vergence from this second surface, is given by:

$$V_2' - V_2 = -\frac{1}{90}$$

Therefore,

$$V_2' = \frac{1}{18} - \frac{1}{90} = \frac{2}{45}$$

But $V_2' = n_w/v'$, where v' is the distance from the concave surface at which the emergent beam cuts the axis and n_w is the refractive index of water. Therefore,

$$v' = \frac{4/3}{2/45} = 30 \text{ cm}$$

Hence, the second principal focus F' in the water is 30 cm beyond the concave surface.

If an incident pencil of rays parallel to the axis is traced from the opposite direction, then $V_1 = 0$, $V_1' = 1/90$ (which is now positive as the incident beam is converged), and

$$V_2 = \frac{1/90}{1-(3/90)/(3/2)} = \frac{1}{88}$$

Also,

$$V_2' = \frac{1}{88} + \frac{1}{20} = \frac{27}{440} = \frac{1}{v}$$

Therefore,

$$v = \frac{440}{27} = 16\tfrac{1}{3} \text{ cm}$$

which is the distance in front of the convex surface of the first principal focus in air.

The power of the optical system is given by Equation 1.20 to be

$$F = F_1 + F_2 - \frac{{}_1d_2}{{}_1n_2} F_1 F_2$$

where F_1 is the power of the first convex surface, F_2 that of the concave surface, ${}_1d_2$ the separation between them, and ${}_1n_2$ the refractive index of the medium between them. Therefore,

$$F = \frac{1}{20} - \frac{1}{90} - \left(\frac{3}{3/2}\right)\left(\frac{1}{90}\right)\left(-\frac{1}{20}\right) = \frac{1}{20} - \frac{1}{90} + \frac{1}{900} = \frac{1}{25}$$

Now the first focal length $f = n/F$, where $n = 1$ for air, and the second focal length $f' = n_w/F$, where $n_w = 4/3$ for water. Hence, $f = 25$ cm and $f' = 25 \times 4/3 = 33$ cm (see Section 1.10).

As the positions of the principal foci relative to the lens surfaces and the principal focal lengths are known, the positions of the principal points and the nodal points are determined, the latter following from consideration of Section 1.9. These locations are shown in Fig. 1.29(b).

A suitable method of experimentally determining the cardinal points would be by means of the nodal slide (Section 1.12.5).

> 5. *Two thin plano-convex lenses of focal lengths (in air) 60·0 and 36·0 cm respectively, are attached coaxially to the ends of a tube 24 cm long and with their plane surfaces facing inwards. If the system is in air, calculate from first principles, and show in a diagram, the positions of its principal foci and unit (principal) points.*
> *When the tube is filled with a transparent liquid and is situated in air, the magnitude of the focal length of the optical system so formed is found to be 27·5 cm. Calculate the refractive index of the liquid.*
> (L. P.)

With the usual notation and referring to Fig. 1.30(a), consider the trace of a parallel pencil of light incident on the lens of focal length 60 cm. Then $V_1 = 0$, $V_1' = 1/60$ and $V_2 = 1/(60-24) = 1/36$; also $V_2' - V_2 = 1/36$, so $V_2' = 1/18$. Therefore v', the distance from the second lens L_2 to the second principal focus F', is 18 cm beyond L_2 (to the right in the diagram).

In the same way, consider an incident pencil from the opposite direction on the lens of focal length 36 cm. Then again, $V_1 = 0$, $V_1' = 1/36$ and $V_2 = 1/(36-24) = 1/12$; also $V_2' - V_2 = 1/60$, so $V_2' = 1/12 + 1/60 = 1/10$. Therefore v, the distance from the first lens L_1 to the first principal focus F, is 10 cm in front of L_1 (to the left in the diagram).

Again, from similar triangles and with reference to Fig. 1.30(a):

$$\frac{H}{h} = \frac{60}{36} = \frac{f'}{18}$$

Therefore, $f' = 30$ cm. Thus, the second focal length is 30 cm and is equal to the first focal length f because the system is in air. The cardinal points are consequently located as shown in Fig. 1.30(b).

If the space between the lenses is filled with a liquid of refractive index n and an incident parallel pencil on the lens of focal length

60 cm is again considered [Fig. 1.30(c)], then $V_1 = 0$ and $V_1' = n/L_1F_1'$. But

$$L_1F_1' = n \times \text{the focal length in air} = 60n$$

Therefore $V_1' = 1/60$ and, from Equation 1.14,

$$V_2 = \frac{V_1'}{1 - V_1'(d/n)} = \frac{1/60}{1 - (1/60)(24/n)} = \frac{n}{60(n - 0 \cdot 4)}$$

The power of the 36 cm lens (which is equal to $1/36$) is given by $V_2' - V_2$. Therefore,

$$V_2' = \frac{n}{60(n - 0 \cdot 4)} + \frac{1}{36} = \frac{4n - 1}{90n - 36}$$

Therefore v', the distance of the second principal focus beyond the second lens, is equal to $(90n - 36)/(4n - 1)$.

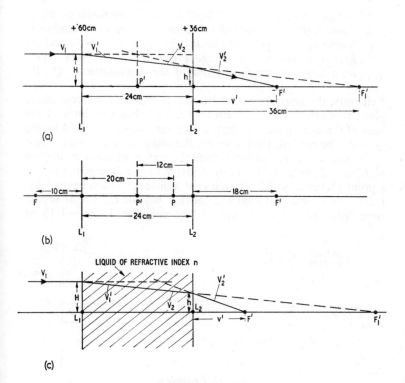

Fig. 1.30. Example 5

By similar triangles, referring to Fig. 1.30(c),

$$\frac{H}{h} = \frac{60n}{60n-24} = \frac{5n}{5n-2} = \frac{f'}{v'} = \frac{f'(4n-1)}{90n-36}$$

where f' is the focal length of the system, given to be 27·5 cm. Thus,

$$\frac{5n}{5n-2} = \frac{27 \cdot 5(4n-1)}{90n-36}$$

Therefore,

$$5n = \frac{27 \cdot 5(4n-1)}{18}$$

or

$$18n = 22n - 5 \cdot 5$$

Hence, the refractive index n of the liquid is equal to $5 \cdot 5/4 =$ 1·375.

6. *A thin converging lens of focal length 20 cm forms an image 1 mm high of a distant object. Determine the nature and focal length of a second lens which, when placed 5 cm behind the first lens, produces an image 1 cm high of the same object on a screen in the focal plane of the combination.* (L. P.)

Evidently, the image of height 1 mm is formed in the second focal plane through F_1' of the 20 cm lens L_1; to obtain an image 1 cm high of the same distant object, a diverging lens L_2 of focal length f_2 would be needed. Furthermore, the image of 1 cm height must be formed in the second focal plane through F' of the combination of L_1 and L_2 (Fig. 1.31). Thus, relative to the divergent lens L_2, a point on the axis at F_1' (which is at a distance 15 cm beyond L_2) must have an image at F' at a distance x' beyond L_2. The divergent cone from F_1' will have an incident vergence $V_1 = -1/15$ at

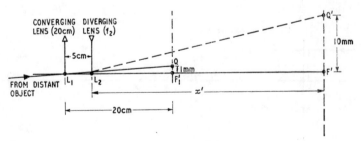

Fig. 1.31. Example 6

L_2, and the emergent divergent cone from L_2 will have an emergent vergence $V_1' = -1/x'$. Since the emergent vergence V_1' minus the incident vergence V_1 equals the power of the divergent lens L_2 (which is $-1/f_2$),

$$-\frac{1}{x'} - \left(\frac{-1}{15}\right) = -\frac{1}{f_2}$$

If Q is the top of the image of 1 mm height formed by L_1 alone and if Q' is the conjugate image of Q (where $Q'F' = 1$ cm), it can be seen from the linear magnification Equation 1.7 that:

$$\frac{x'}{15} = 10$$

Therefore $x' = 150$ cm. Hence,

$$\frac{1}{f_2} = \frac{1}{150} - \frac{1}{15} = -\frac{9}{150}$$

Therefore f_2, the focal length of the diverging lens, is $-16\frac{2}{3}$ cm.

7. *A telephoto lens consists of a converging lens and a diverging lens of focal lengths 5·0 cm and 2·0 cm respectively placed coaxially 3·5 cm apart. At what distance from the diverging lens will the image of a distant object be formed and what will be the size of this image if the object subtends an angle of 5°?* (L. G.)

A pencil of rays parallel to the axis incident upon the 5 cm converging lens L_1 [Fig. 1.32(a)] is traced through the system. With the usual notation, $V_1 = 0$, $V_1' = 1/5$ and $V_2 = 1/(5-3\cdot5) = 2/3$; also $V_2' - V_2 = -1/2$, so $V_2' = 2/3 - 1/2 = 1/6$. Therefore v', the distance of the second principal focus F' beyond the diverging lens L_2, is 6 cm, and the image of a distant object will be formed in the second focal plane passing through this point.

By similar triangles in Fig. 1.32,

$$\frac{H}{h} = \frac{5}{1\cdot5} = \frac{f'}{6}$$

Therefore, the second focal length $f' = 20$ cm. The length y of the image will thus be given by [from Fig. 1.32(b)]:

$$y = f' \tan 5° = 20 \tan 5° = 1\cdot7 \text{ cm}$$

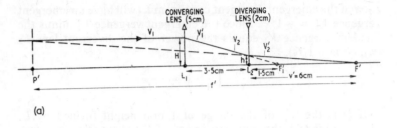

(a)

(b)

Fig. 1.32. Example 7

8. *A thin converging lens of focal length 4 cm and diameter 5 cm has a 3 cm diameter stop placed 2 cm in front of it. Calculate the position and size of the exit pupil and draw on a diagram roughly to scale the marginal and chief rays from a point 8 cm in front of the lens and 1·5 cm off the axis to the corresponding image point.* (L. P.)

The stop will evidently be the aperture stop of the system (Section 1.13) for all object points such as Q on the axis of the system at a distance greater than x from the lens, where

$$\frac{x}{2\cdot5} = \frac{x-2}{1\cdot5}$$

Therefore $x = 5$ cm, as is seen from Fig. 1.33(b).

The entrance pupil is the image of the aperture stop in all optical elements before it (Section 1.13), and thus the stop is also the entrance pupil of the system because no optical elements precede it. The exit pupil is the image of the stop in all optical elements succeeding it, i.e. in the lens of focal length 4 cm. It will therefore be at E' at a distance LE' from the lens L, E' being a point on the axis which is an image of E (an axial object point at the centre of the aperture stop [Fig. 1.33(a)].

The vergence V' of a divergent cone of rays from E incident on to lens L is $-1/2$; the emergent vergence V_1' of the cone leaving L is therefore given by:

$$V_1' - V' = \text{power of lens } L = \frac{1}{4}$$

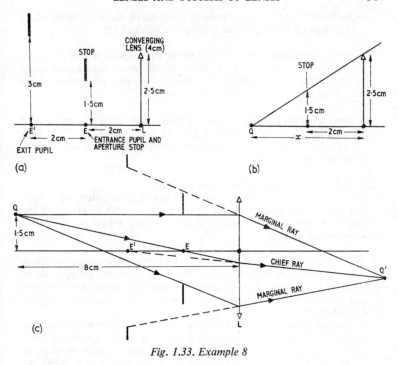

Fig. 1.33. Example 8

Therefore,

$$V'_1 + \frac{1}{2} = \frac{1}{4}$$

Therefore, $V'_1 = -1/4$; thus a divergent cone of rays leaves L and must come from a virtual image point 4 cm in front of L. Hence the exit pupil is 4 cm in front of L, i.e. to the left-hand side of L in Fig. 1.33(a). The linear magnification m is given by:

$$m = \frac{\text{diameter of exit pupil}}{\text{diameter of entrance pupil}} = \frac{4}{2} = 2$$

Hence, the diameter of the exit pupil is 6 cm. The diagram required in the last part of this question is shown in Fig. 1.33(c).

Exercise 1

1. Light from a point P in air is converged by a spherical refracting surface to a point Q in glass. The distance $PQ = 18$ cm and the point P is twice

as far from the surface as the point Q. Find the radius of curvature of the surface given that the glass has a refractive index of 1·5.

2. A point object in air is at a distance of 100 cm from the pole of a concave spherical refracting surface of radius 17 cm separating air from glass. A virtual image of this object is formed at a distance of 40 cm from the pole. Find the refractive index of the glass.

3. A thin biconcave lens has spherical surfaces with radii of curvature of 10 cm and 15 cm and is made of glass of refractive index 1·5. If light diverges from a point on the axis 20 cm from the lens, where will the conjugate image be formed?

4. Define the cardinal points of a lens system.
   Derive expressions for the positions of the cardinal points of a thick bispherical lens in air and hence or otherwise find the positions of the cardinal points of a solid sphere (in air) of radius 5·0 cm of material of refractive index 1·62.   (L. G.)

5. Define the term focal length for a coaxial system of spherical refracting surfaces. Obtain an expression for that of a pair of thin lenses in air in terms of the powers of the components and their separation. Describe and explain a method whereby the power of a system can be determined experimentally.
   A thin convex lens of power 5 dioptres forms an image 0·40 cm high of a distant object. Find the nature and power of a second lens which when placed 5 cm on the image side of the first lens produces in the focal plane of the system a real image 1·0 cm high of the same distant object. At what distance from the second lens will this image be formed?   (L. G.)

6. Two thin converging lenses of focal lengths 3·0 and 1·0 cm are placed 2·0 cm apart to form an eyepiece for a microscope. Assuming formulae for thin lenses only, show that there is a real focal point at 0·5 cm outside the more powerful lens and determine the position of the corresponding principal (or unit) plane and the focal length of the system. Describe a method for measuring this focal length.   (L. Anc.)

7. State, and illustrate by diagrams, the properties associated with principal planes, nodal points, and principal foci of a system of coaxial lenses.
   Two coaxial converging lenses each of focal length 20 cm are 10 cm apart. Find the positions of (a) the principal foci, and (b) the principal planes of the system.   (L. Anc.)

8. Two homogenous media having different refractive indices are separated by a thin glass lens. From first principles, derive a relationship between the distances of the two conjugate points from the lens in terms of the radii of curvature of the surface of the lens, the refractive index of the glass, and the refractive indices of the two media.
   The meniscus at the top of a column of water in a measuring cylinder has a radius of curvature equal to the height of the column. Calculate the

ratio of the real and apparent depths of the bottom of the cylinder. Refractive index of water $= 1.33$.   (L. Anc.)

9. Define the cardinal points of a thick lens, and obtain the positions of these for a glass sphere in air.
   A sphere of radius 5 cm and a hemisphere of radius 15 cm both made of glass of refractive index 1·50 are arranged with the pole of the curved surface of the hemisphere in contact with the sphere, and a parallel beam of light is directed normally on to the plane face of the hemisphere. Assuming that a stop restricts the beam to the paraxial region, calculate the position of the final focus.   (L. G.)

10. Describe how the cardinal points of a coaxial lens system may be located experimentally. Two thin converging lenses of focal lengths 5 and 10 cm are placed coaxially 10 cm apart in air. Draw the system to scale showing the calculated positions of its cardinal points.   (L. Anc.)

11. Derive an expression for the distance apart of the principal points of a lens system consisting of two thin converging coaxial lenses of focal lengths f_1 and f_2 at a distance z apart in air. Taking $f_1 + f_2 = 1$ plot a rough graph of the variation of this distance with z and calculate the value for which it is a minimum.   (L. P.)

12. A converging mensicus lens, 3 cm thick, has its concave surface in contact with water of refractive index 4/3 and its convex surface exposed to air. If the material of the lens has a refractive index 3/2 and the radii of curvature of the faces are 10 cm and 15 cm respectively, determine the positions of the cardinal points of the system. Illustrate your answer by a diagram drawn roughly to scale, and indicate briefly how you would locate the cardinal points experimentally.   (L. P.)

13. Define the terms *focal length* and *principal plane* as applied to a coaxial system of lenses.
    An object in a closed chamber is to be viewed through a thin window placed 10 cm from the object. A microscope objective outside the chamber consists of a converging lens, focal length 5 cm, in contact with this window, and a diverging lens, 7 cm from the first lens, and of 4 cm focal length. Find the cardinal points of the system, the plane of the image and its magnification, and give a ray diagram.   (L. P.)

14. Describe and give the theory of a method of determining the cardinal points of a thick lens. Calculate the position of these points for a system consisting of a glass hemisphere of radius 10 cm with water in contact with the plane side, and mark them on a diagram drawn approximately to scale. (Assume the refractive indices of glass and water to be 3/2 and 4/3 respectively.)   (L. P.)

15. Define the principal points of a lens system and derive an expression for the focal length of a combination of two thin lenses in air in terms of

their focal lengths and distance apart. How would you determine the focal length of such a combination experimentally?

A thin converging lens of focal length 20 cm forms an image 4 mm high of a distant object. Determine the nature and focal length of a second thin lens which, when placed 5 cm behind the first lens, produces an image 1 cm high of the same object on a screen in the focal plane of the combination. (L. P.)

16. Write an account of the stops used in coaxial optical systems, include definitions, with appropriate diagrams, of the terms aperture stop, entrance pupil, exit pupil, field stop, entrance and exit windows, and telecentric systems.

17. What are meant by: (i) the cardinal points, and (ii) the focal length of a lens system?

Calculate the focal length of a system consisting of a thin converging lens, focal length 5 cm, coaxial with a glass hemisphere, radius 3 cm, when the lens is placed 2 cm from the curved surface of the hemisphere. The refractive index of the glass may be taken to be 1·5. (G. I. P.)

18. An equi-convex lens of refractive index 1·50 has a focal length in air of 20 cm. It is used as a window for a long fish-tank of water of refractive index 1·33 so that there is water on one side of the lens and air on the other.

Find the focal points of the system and illustrate your conclusions by preparing a scale drawing. (G. I. P.)

19. A line object $\frac{1}{2}$ inch long is perpendicular to the axis of a converging lens of focal length 3 inches and diameter 1 inch, and has one end on the axis at a point 1 inch to the left of the lens. A second converging lens of focal length 1 inch and diameter 2 inches is placed coaxially to the right of the first in such a position that all the rays from the object transmitted by the first lens just pass through the second.

Draw a full-scale diagram of the system distinguishing clearly between constructional and actual ray paths. Determine graphically the distance between the two lenses, and the position and size of the final image. Compare your result with that obtained by calculation. (Assume for the purpose of this question that the lenses are thin, and that the paraxial formulae apply.) (G. I. P.)

20. A compound lens is made from two identical equiconvex thin lenses of refractive index 1·5 and focal length 5 cm fitted at the ends of a tube 6 cm long. Show that the power of the lens is the same when the space between the lenses is filled with a liquid of refractive index 1·5 as it is when the tube contains air only.

Find the position of the cardinal points in each case and hence show that the distance between the object and the image for any specified magnification is 10 cm greater when the tube is filled with liquid than when it contains none. (Aston)

21. Derive a formula for the positions of the principal planes of a system consisting of two thin lenses.

A convex lens of focal length 20 cm is placed in air, 12 cm from a concave lens of 10 cm focal length. Find the focal length of the combination, and the positions of the focal points and principal planes. (M. P.)

CHAPTER 2

Aberrations of Optical Components and Systems

2.1 The Aberrations of Optical Systems Used with Monochromatic Light

If an optical system is traversed by monochromatic light (ideally light of a single wavelength, but in practice with a spread of wavelength from λ to $\lambda + \Delta\lambda$ where $\Delta\lambda << \lambda$), chromatic aberration (Section 2.15) due to dispersion (caused by variation of the refractive index with wavelength) is absent. Ideally, if such a system is observed monochromatically, it should give a point image conjugate to a point object, a line image for a line object, and a plane for a plane.

Most practical systems consist of coaxial spherical refracting or reflecting surfaces, or combinations of these. Such surfaces will deflect the rays by refraction or reflection from objects (point, line or plane) in a manner depending on the object position and will give rise to images which are not corresponding points, lines or planes.

The nearest approach to the ideal optical system is realised with very narrow cones of rays passing through the centre of the system (i.e. with paraxial cones of rays). The monochromatic aberrations which result with non-paraxial (called extra-paraxial or extra-axial) cones—and present, even if small, with paraxial cones—were first investigated thoroughly by von Seidel. He classified what are now called the five *Seidel aberrations*. These are spherical aberration, coma, astigmatism, curvature of the field, and distortion.

2.2 Spherical Aberration

A light ray which passes from an object point to a conjugate image point in a given medium will traverse an optical path length equal to the actual geometrical path length multiplied by the refractive index of the medium. If different media 1, 2, 3, ... of respective refractive indices n_1, n_2, n_3, ... are successively encountered in this path, and if the corresponding actual path lengths in these media are l_1, l_2, l_3, ..., then the optical path length traversed by the light ray is $n_1l_1 + n_2l_2 + n_3l_3 + \ldots$ or, in general, Σnl.

A reflecting or refracting surface may be so shaped that the rays from a specific object point to a specific conjugate image point via this surface follow optical paths which are all equal, independently of the points on the surface where the reflection of refraction occurs. Such rays will all have the same transit time, and the surface is said to be *aplanatic*. An aplanatic surface is capable of producing a close approximation to a point image of a point object; more specifically, the image will be free of spherical aberration and of coma. In general, the shapes of aplanatic surfaces will not be spherical.

An example of an aplanatic concave reflecting surface is an ellipsoid of revolution, obtained by the rotation of an ellipse about the axis through its foci. An object point at one focus P of this ellipsoid will produce an image, free from spherical aberration and from coma, at the other focus Q (Fig. 2.1). This is because of the two well-known properties of ellipses: that the path lengths from P to Q via any point C on the ellipse are independent of the location of C; and that PC and CQ make equal angles to the

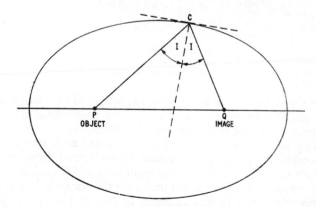

Fig. 2.1. An aplanatic concave reflecting surface

normal to the ellipse at *C*. Note, however, that this system is aplanatic only for the specific object and image points *P* and *Q*, and not for any other points on the axis through *PQ*.

If an aplanatic surface is replaced by a spherical one, the two surfaces will only coincide closely for a limited cone which is reflected or refracted at the central region of the spherical surface. It will be only for this cone that rays from a point object to the spherical surface will give rise to well-formed point images. This central region of near coincidence is the paraxial region, which may be defined as that region in which first-order theory is valid; so $\sin \theta = \theta$ and $\cos \theta = 1$, where θ is the semi-angle of the cone of rays concerned. Outside this region, any given zone of the spherical surface will form an axial image displaced from the paraxial image. It is this defect, arising from the spherical shape of the surface, which is called spherical aberration; an example is shown in Fig. 2.2, where image-forming rays after refraction or reflection are illustrated.

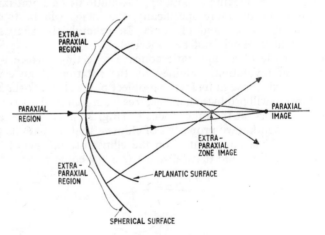

Fig. 2.2. Spherical aberration

For a given extra-paraxial zone, the spherical aberration may be measured geometrically in terms of the *longitudinal spherical aberration* λ and the *transverse spherical aberration* τ. The former, λ, is decided by the axial displacement of the zone image from the paraxial image; and τ is decided by the lateral displacement from the paraxial image of a ray from the zone (Fig. 2.3).

A continuous range of zones including the paraxial region will thus give rise, in a typical example, to rays in the image space such as those shown in Fig. 2.4. To avoid a complex diagram,

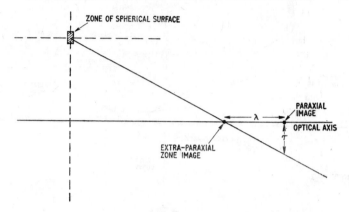

Fig. 2.3. Longitudinal and transverse spherical aberration

rays from only three zones are shown above the optical axis
and from only the extreme zone below this axis; in practice,
there will be an infinite number of zones. Rays from the extra-
paraxial zones will intersect before reaching the optical axis as,
for example, at *A*. These points of intersection will occur on a
curve in the plane of the diagram called the *caustic curve*, which
has a *cusp* at *Q*. An everyday example is the bright line of light
on the surface of a cup filled with liquid: this is the caustic
curve delineated in the surface of the liquid and is due to intersections between rays of light reflected from the cylindrical wall
of the cup. In Fig. 2.4, only the lower half of the caustic curve
is shown; the upper half, above the optical axis, is omitted for
clarity.

If a screen is placed at position 1 shown in Fig. 2.4 (i.e. in a
plane perpendicular to the optical axis and including the paraxial
image *Q*), the image seen will have a bright centre but will be
surrounded by a diffuse circular patch or halo of fainter light
due to extra-paraxial rays; this is illustrated by image 1 of Fig. 2.4.
If the screen is moved nearer to the optical element having a
range of zones (i.e. to position 3, which is decided by the plane
of the first intersections between extreme extra-paraxial rays),
it will show an image with a bright circular edge of outer diameter *AB* with fainter light within; see image 3 of Fig. 2.4. In
between the extreme positions 1 and 3, there is a position 2 at
which the image discerned is most uniformly illuminated and
has its least diameter *CD*. This position gives the best approach
to an image of a luminous point object formed by the optical
element; it is called the *circle of least confusion*.

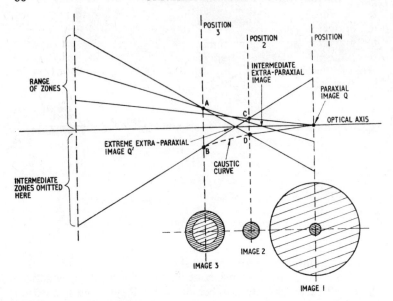

Fig. 2.4. The consequences of spherical aberration showing the caustic curve and the nature of the images of a luminous point object

If an optical system is free from longitudinal spherical aberration λ and transverse spherical aberration τ (Fig. 2.3), it follows, irrespective of the zone concerned, that an object of finite extent at a distance x from the first principal plane will give rise to an image at a distance x' from the second principal plane and not one which is spread over a distance from x' to $x' - \lambda$. The linear magnification $m = x'/x$ is, therefore, constant irrespective of the zone. In other words, to be completely free from spherical aberration, all the zones of a given optical system must have the same power for given object and conjugate image positions.

2.3 Methods of Reducing Spherical Aberration

There are four ways of reducing spherical aberration.

2.3.1 LIMITING THE APERTURE OF THE OPTICAL SYSTEM

One obvious and frequently used remedy is to limit the aperture of the optical system to a region giving tolerable spherical aberration. This will inevitably reduce the amount of light that traverses the optical system from a given luminous object and so will result in fainter images. Reduction of the aperture also reduces the resolving power of the system (Section 12.2), though it does have the advantage of increasing the depth of field and of focus (Section 3.8).

2.3.2 USING APLANATIC SURFACES WITH A GIVEN OBJECT AND CONJUGATE IMAGE POSITION

This method has been considered briefly in Section 2.2. Unfortunately, the fabrication by grinding and polishing of aspherical surfaces is more difficult and costly than of spherical ones. Aspherical aplanatic concave reflectors are used: for example, a paraboloidal mirror with a luminous source at its focus forms a collimator. In special instances, a spherical surface may be aplanatic; such a surface is used in the design of oil-immersion microscope objectives (Section 3.6). Thus, if a system is designed with common aplanatic points for successive aplanatic surfaces, it is, in theory, free from spherical aberration. This system, if refracting, is limited by the fact that one of the aplanatic points must be virtual.

2.3.3 USING A SCHMIDT CORRECTOR PLATE

A Schmidt corrector plate can be used to correct the spherical aberration of a spherical concave mirror. This ingenious solution, in which a specially figured transparent plate of glass or plastic is used to balance out the spherical aberration due to a large aperture concave mirror, was developed for reflecting telescopes (Section 3.4); it has also been used, for example, to project onto a screen an enlarged image of the picture obtained on a television receiver cathode ray tube.

2.3.4 CHOOSING AND ARRANGING THE SPHERICAL SURFACES IN AN OPTICAL SYSTEM

The spherical surfaces in an optical system can be chosen and arranged in such a manner that the aberration due to the extra-paraxial zones is minimised. This is the method most widely used in the design of systems; some aspects of the procedure are considered in the following sections.

2.4 The Condition that a Thin Lens should have Minimum Spherical Aberration

A point object is placed on the optical axis of a thin lens at a distance x from the centre of this lens. The incident vergence will be n/x, where n is the refractive index of the object space medium. If x' is the distance along the optical axis to the paraxial image, the emergent vergence will be $V' = n'/x'$, where n' is the

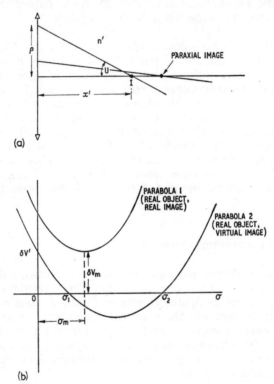

Fig. 2.5. (a) Convergent cone of rays from an extra-paraxial zone of a thin lens; (b) change of emergent vergence $\delta V'$, plotted against the lens shape factor, σ, for a thin lens and two different values of the parameter ϕ

refractive index of the image space medium. Consider an extra-paraxial zone of this lens lying between radii ϱ and $\varrho + \delta\varrho$. Owing to longitudinal spherical aberration, the image distance x' will change to an extent depending on ϱ^2.

In Fig. 2.5(a), let U be the semi-angle at the vertex of the cone of rays from the extra-paraxial zone of radius ϱ, the image point at this vertex being I. Instead of assuming that $\sin U = U$, as would hold for a paraxial cone of rays, the next term in the series expressing $\sin U$ is taken into account, so that

$$\sin U = U - \frac{U^3}{3!}$$

Then it can be shown that, as ϱ changes from ϱ to $\varrho + \delta\varrho$, the emergent vergence V' will change by an amount

$$\delta V' = \frac{\varrho^2}{(f')^3}(A\sigma^2 + B\sigma\phi + C\phi^2 + D) \tag{2.1}$$

where $\sigma = (r_1 + r_2)/(r_1 - r_2)$, and is known as the *lens shape factor*, r_1 and r_2 being the radii of curvature of the surfaces of the thin lens; $\phi = (x + x')/(x - x')$ and is known as the *lens position factor*; f' is the second focal length of the lens for the particular zone, or the *zone focal length*; and A, B, C and D are functions of the refractive indices of the materials of the lens and the media only.

The plot of $\delta V'$ against σ for given values of ϱ, f', A, B, C and D is a parabola for each given value of the lens position factor ϕ. A family of parabolae will be obtained for various values of ϕ as the parameter. Two parabolae are shown in Fig. 2.5(b): parabola 1, where ϕ is such that the parabola is entirely above the σ-axis; parabola 2, where ϕ is such that the parabola intersects the σ-axis at two values of σ.

A parabola of the the first type corresponds to a situation where a real image is formed of a real object; thus, $\delta V'$ cannot be zero, so the spherical aberration cannot be zero. The important consideration is therefore the minimum value of $\delta V' = \delta V_m$ at the apex of the parabola. This occurs for a value σ_m of the lens shape factor σ. Hence, minimum spherical aberration is achieved by shaping the lens surfaces to give the correct values of r_1 and r_2. The spherical aberration is then a minimum for a given zone within the lens aperture; since this condition is independent of the zone radius ϱ, this aberration will be a minimum for all zones. A lens so shaped is called a 'crossed' lens. It can be shown that the bending of the light rays passing through a crossed lens is equally distributed between the lens surfaces. For a biconvex lens of glass of refractive index $1\cdot5$, this equality of deviations is achieved when the radii of curvature of the surfaces are in the ratio $1:6$, providing the more convex surface faces the incident light. Furthermore, the larger the refractive index of the lens material, the smaller σ_m can be made.

A parabola of the second type can only be obtained if the image of a real object is virtual. Now $\delta V'$ is zero at the two values

σ_1 and σ_2 of σ, and the spherical aberration is zero. From Equation 2.1, it then follows that

$$A\sigma^2 + B\sigma\phi + C\phi^2 + D = 0$$

and

$$\left.\begin{array}{c}\sigma_1 \\ \sigma_2\end{array}\right\} = \frac{-B\phi \pm \sqrt{[B^2\phi^2 - 4A(C\phi^2 + D)]}}{2A}$$

In 1830, Lister showed that a combination of two thin lenses in contact, one converging and the other diverging, could be designed to give minimum spherical aberration for two pairs of conjugate points, one a real pair and the other pair comprising one real and one virtual point. These points for such a doublet are usually called the *Lister points;* they are of use in the design of microscope objectives (Section 3.6).

2.5 The General Condition for Freedom from Spherical Aberration: the Abbe Sine Condition

If a given optical system is to be completely free from spherical aberration, all zones must have the same power; or, for given object and conjugate image positions, the lateral magnification must be the same for all zones (Section 2.2).

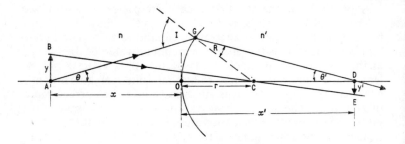

Fig. 2.6. The Abbe sine condition

In Fig. 2.6, A and D are two axial points conjugate with respect to a spherical surface of radius r and centre of curvature C; the pole of the surface is at O, which separates two media of refractive indices n and n'. $AO = x$ and $OD = x'$. An object AB of height y at A gives rise to an image DE of height y'. If AB and DE are small, the effects of astigmatism (Section 2.11) and distortion (Section 2.13) may be neglected.

A ray from B normal to the spherical surface is undeviated by refraction and passes through C, the centre of curvature, to E. If m is the lateral magnification of AB,

$$m = \frac{y'}{y} = \frac{CD}{AC}$$

because triangles ABC and DEC are similar.

Consider a point G on the spherical surface in any zone (paraxial or extra-paraxial). A ray from A to G makes an angle θ with the optical axis and is incident at an angle I at G; after refraction, the path of this ray is GD, the angle of refraction being R, and the angle between GD and the optical axis being θ'. In the triangle ACG,

$$\frac{AC}{\sin I} = \frac{GC}{\sin \theta} = \frac{r}{\sin \theta}$$

Similarly, for the triangle DCG,

$$\frac{CD}{\sin R} = \frac{r}{\sin \theta'}$$

Therefore,

$$m = \frac{CD}{AC} = \frac{\sin R \sin \theta}{\sin I \sin \theta'} = \frac{n \sin \theta}{n' \sin \theta'}$$

If m is to be constant for all rays from A,

$$\frac{\sin \theta}{\sin \theta'} = \text{constant} \tag{2.2}$$

as n/n' is constant.

This result is known as the *Abbe sine condition*. If no other aberrations are present, ray-tracing procedures may be used to obtain values of θ and θ'. A plot of $\sin \theta$ against $\sin \theta'$ will give a measure of the spherical aberration present in a given system.

2.6 Coma

Coma is a complex aberration occurring with non-axial cones of rays, i.e. cones of which the chief rays are at angles to the optical axis. It is prevalent when an extended object is used such that light rays emanate in a cone from points in the object which are not on the optical axis of the system. Coma arises, as does spherical aberration, because of the variation in optical power with zone radius; but this is now combined with the oblique incidence of various rays in the non-axial cone to a given zone of the entrance pupil of a system.

Coma is most easily illustrated for a thin lens or for the thin lens equivalent to a coaxial optical system. In Fig. 2.7(a), suppose an axial point object at P has a conjugate axial point image

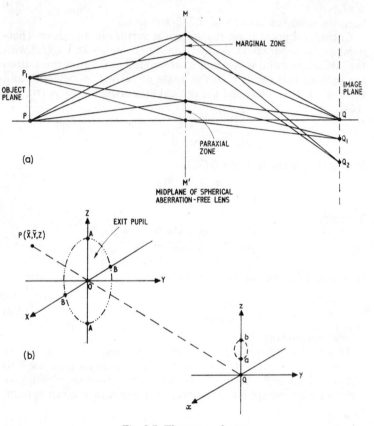

Fig. 2.7. The cause of coma

at Q formed by a lens having a midplane MM' and which is assumed to be free of spherical aberration. All rays from P are focused at Q. However, if coma exists, rays from a non-axial point P_1 (which is on a plane surface through P perpendicular to the optical axis) are not all focused at a single point on a conjugate plane through Q; instead, there is a series of circular patches of light between such points as Q_1 and Q_2, where Q_1 is formed by a cone of rays through the central zone of the lens, and Q_2 is formed by a cone through a marginal zone of the lens.

For a thin lens, the exit and entrance pupils coincide. To illustrate coma more fully, suppose the exit (or entrance) pupil lies in the plane XOZ of the Cartesian set of coordinates denoted by the capital letters X, Y and Z; the image positions are referred to a separate set of parallel coordinates denoted by the small letters x, y and z [Fig. 2.7(b)]. Let the dotted circle $ABAB$ of radius ϱ in the plane XOZ represent a zone of the exit pupil. A non-axial point object P having coordinates (X, Y, Z) (where the bar over the letter represents a negative value in accordance with the usual Cartesian coordinate convention) gives rise to a paraxial image point Q. For convenience, Q will be taken as the

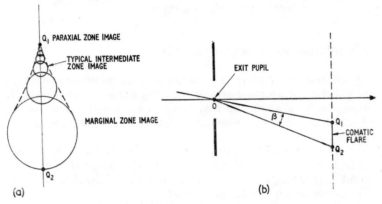

Fig. 2.8. *Appearance of the image of a non-axial point object formed by a lens having coma: (a) flare-shaped image (exaggerated); (b) flare-shaped image extends from Q_1 to Q_2*

origin of the coordinates x, y and z. Q will have coordinates $(X, Y, \bar{Z})$ relative to point O in the example shown in Fig. 2.7(b). The chief ray of the paraxial cone from P will pass through O, the centre of the zone $ABAB$, to reach Q, but Q will not be on the optical axis of the system. Rays leaving the exit pupil zone of radius ϱ from two diametrically opposite points, such as AA on the Z-axis, will intersect at point a in the z-axis — either above or below Q; thus, the point a will have coordinates $(0, 0, z)$ or $(0, 0, \bar{z})$.

It can be shown that rays from any other points in the exit pupil zone which are at opposite ends of a diameter situated at an angle θ to AA will intersect at a point in the plane xoz at a position which is rotated by 2θ from the point a. Therefore rays leaving the exit pupil zone from points BB at $\theta = 90°$ to AA will intersect at a point b rotated $180°$ from a. It follows that the whole cone of rays leaving $ABAB$ results in a circle in the plane xoz with a diameter ab. Thus, a given zone will image a non-axial point object as a *comatic circle*, of which the diameter will increase with increase of the zone radius. Furthermore, the centre of this comatic circle will be separated from the paraxial image point Q by a distance which increases with the zone radius ϱ.

The continuous set of concentric zones in the exit pupil will consequently image a non-axial point object in the form of a series of overlapping circular images, the diameter of these images decreasing towards the paraxial zone image [Fig. 2.8(a)]. The characteristic 'flare' or 'comet' shaped image has given rise to the term 'coma' for this aberration.

2.7 Reduction of Coma of a Thin Lens

As with spherical aberration, coma may be minimised by stopping down the aperture and so reducing the entrance (also exit) pupil, or by shaping the lens surfaces to non-spherical forms. A thin lens which has spherical surfaces can also be 'crossed' for coma.

A convenient measure of coma is in terms of the angle β [Fig. 2.8(b)] between the emergent rays from the centre O of the exit pupil to Q_1 and Q_2, points at the extremities of the comatic flare. For a thin lens it can be shown that

$$\tan \beta = A'\sigma + B'\phi \tag{2.3}$$

where $\sigma = (r_1 + r_2)/(r_1 - r_2)$ is the lens shape factor, as in Section 2.4; and $\phi = (x + x')/(x - x')$ is the lens position factor, where x and x' are measured from the lens centre to the object and paraxial image points respectively. A' and B' are functions of the refractive indices of the lens material and the media in which it is situated.

From Equation 2.3, it is clear that values of σ and ϕ can always be found to make β equal to zero. Hence, for a given object and paraxial image (i.e. for given x and x'), a thin lens may be shaped so that its spherical surfaces have particular values of r_1 and r_2 to give zero coma. Such a lens is said to be 'crossed' for coma. In practice, a lens 'crossed' for spherical aberration is also very nearly 'crossed' for coma.

2.8 Coma and the Abbe Sine Condition

If the Abbe sine condition (Equation 2.2) is satisfied for an optical system, the lateral magnification will be the same for every zone. If the system is free from all other aberrations, this sine condition will, therefore, give a measure of its coma.

2.8.1 SPHERICAL REFRACTING APLANATIC SURFACES

In the absence of other correcting factors for the aberrations not so far considered (astigmatism, curvature of the field, and distortion), a spherical refracting surface which satisfies the sine condition for a given pair of conjugate points will be free of both spherical aberration and coma for axial objects of small extent. As already stated (Section 2.4) one of these points will be real and one virtual. It is useful to consider the positions of these points in certain cases of practical value.

2.9 The Aplanatic Points of a Single Spherical Refracting Surface

In Fig. 2.9, P is a small object within a spherical surface of radius r containing a medium of refractive index n and surrounded by a medium of refractive index n'. The obvious example is a sphere of glass in air. P is at a distance x from the centre O of the sphere. The image formed at Q outside the sphere in the medium of refractive index n' is at a distance x' from O. From symmetry considerations, it is clear that the image at Q is on the line OP produced.

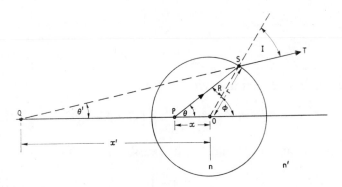

Fig. 2.9. The aplanatic points of a sphere

A ray PS at an angle θ to the axis OP meets the spherical surface at an angle R to the normal and gives rise to a ray ST in the surrounding medium at an angle I to this same normal. Extension of ST backwards to the axis locates the image Q. With the angles θ, θ' and ϕ as marked on Fig. 2.9, in triangle PSO:

$$\frac{x}{\sin R} = \frac{PS}{\sin \phi}$$

Similarly, in the triangle QSO:

$$\frac{x'}{\sin I} = \frac{QS}{\sin \phi}$$

Therefore,

$$\frac{PS}{QS} = \frac{x \sin I}{x' \sin R} = \frac{xn}{x'n'} \tag{2.4}$$

Now,

$$PS^2 = x^2 + r^2 + 2xr \cos \phi$$

and

$$QS^2 = (x')^2 + r^2 + 2x'r \cos \phi$$

Therefore,

$$\frac{PS^2}{QS^2} = \frac{x^2 + r^2 + 2xr \cos \phi}{(x')^2 + r^2 + 2x'r \cos \phi} = \frac{x^2 n^2}{(x')^2 (n')^2} \tag{2.5}$$

If P and Q are to be aplanatic points with respect to the spherical surface, Equation 2.5 must be true not only for the particular point S on the sphere illustrated in Fig. 2.9 but for *any* position of S on this sphere (i.e. for all values of the angle ϕ). For a wide cone of rays from the small object P, there will then be a small image at Q free from spherical aberration and coma. Hence, equating coefficients of $\cos \phi$ in Equation 2.5,

$$(x')^2(n')^2 x = x^2 n^2 x'$$

or

$$\frac{x'}{x} = \left(\frac{n}{n'}\right)^2 \tag{2.6}$$

Equating terms not involving $\cos \phi$:

$$(x')^2(x^2 + r^2) = \left(\frac{n}{n'}\right)^2 \left[x^2(x')^2 + x^2 r^2\right] \tag{2.7}$$

From Equations 2.6 and 2.7, it is easily shown that

$$x = \left(\frac{n'}{n}\right) r \tag{2.8}$$

and

$$x' = \left(\frac{n}{n'}\right) r \tag{2.9}$$

Note that the Abbe sine condition (Section 2.5) is satisfied because, in triangle QPS,

$$\frac{\sin \theta}{\sin \theta'} = \frac{QS}{PS} = \frac{x' \sin R}{x \sin I} = \frac{x'n'}{xn}$$

from Equation 2.4. As $x'/x = (n/n')^2$ from Equation 2.6,

$$\frac{\sin \theta}{\sin \theta'} = \left(\frac{n}{n'}\right)^2 \left(\frac{n'}{n}\right) = \frac{n}{n'} = \text{constant}$$

To summarise, it has been shown that, if a sphere of material of refractive index n is within a medium of refractive index n', two aplanatic points exist at distances $n'r/n$ and nr/n' from the centre of the sphere. A small object at one of these points gives rise to an image, free from spherical aberration and coma, at the other point. This result is valuable in the design of oil-immersion objectives for microscopes (Section 3.6).

2.10 The Aplanatic Points of a Thin Spherical Lens

In order to find the aplanatic points of a thin lens with a given shape factor $(r_1+r_2)/(r_1-r_2)$, it would be necessary to satisfy Equations 2.1 and 2.3 for a given position factor $(x+x')/(x-x')$. This is not generally possible; indeed, it can be shown that only one solution apart from $(r_1+r_2)/(r_1-r_2) = (x+x')/(x-x') = 0$ can be found.

This solution is obtained by comparison with a lens in the form of a sphere (Section 2.9). The small object on the axis must be at an aplanatic point a distance $n'r/n$ from the centre of a sphere of refractive index n' within a medium of refractive index n (Equation 2.8). Suppose a thin meniscus lens of refractive index n, with surfaces 1 and 2 of radii r_1 and r_2 respectively (Fig. 2.10), is used. O is the centre of curvature of surface 2, and P is the centre of curvature of surface 1. A small object on the axis at P, which is in air or the medium of refractive index n' and at a distance r_1 from surface 1, will produce a ray such as PS which is normal to surface 1 at S. Thus the ray PS will enter the lens undeviated to meet the surface 2, where it will undergo refraction to appear to come from an image at Q. In effect, because of the lack of refraction at S, P is imbedded in a sphere of radius

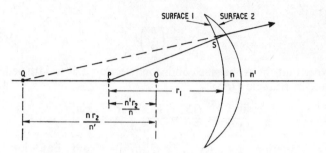

Fig. 2.10. The aplanatic points of a thin meniscus lens

r_2 and refractive index n. If, furthermore, P is situated so that $PO = n'r_2/n$ then P is an aplanatic point, and the image at Q is at the second aplanatic point at a distance nr_2/n' from O.

2.11 Astigmatism

In spherical aberration, the focusing of cones of rays symmetrical about the optical axis was considered; and coma is concerned with the focusing of skew cones, of which the chief rays do not lie along the optical axis. Even if an optical system is free from both these aberrations, it may still focus fans of rays to two different points. By a fan of rays is meant a 'two-dimensional' cone, or rays in a plane which diverge or converge. This effect, in general, is due to the combined effect of oblique incidence of the rays and the variation of power of the optical system for different fans of rays. An extreme example is a cylindrical lens which has zero power for certain fans in one plane and does not focus them at all, whereas for fans in a perpendicular plane the maximum focusing effect is obtained.

In Fig. 2.11, suppose the exit pupil of an optical system lies in the plane yOz perpendicular to the optical axis Ox, where O is the centre of this pupil, and x, y and z are a Cartesian set of coordinates. Within a circular zone of this exit pupil, consider the square $abcd$, of which the sides ab and cd are parallel to the y-axis, whilst ad and bc are parallel to the z-axis. Let P be a point on the object (but off the optical axis) having coordinates $(\bar{x}_1, 0, \bar{z}_1)$, where $\bar{x}_1$ and $\bar{z}_1$ are particular negative values of x and z respectively. The paraxial image of P is at Q and has coordinates $(x_2, 0, z_2)$. Thus, OQ is the chief ray on a paraxial cone of rays at an angle to the optical axis Ox, both OQ and Ox being in the xOz plane.

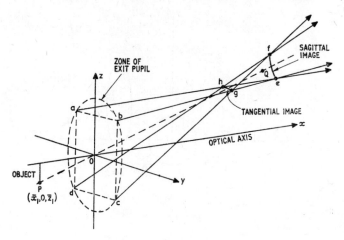

Fig. 2.11. Astigmatism

The fan of rays from points *ab* intersects at *e*, whereas that from *cd* intersects at *f*. In addition, the fan from *ad* intersects at *h*, and that from *bc* intersects at *g*. When all fans leaving the zone are considered, two short orthogonal curves *hg* and *fe*, forming the image of the object point, are produced. The pencil of rays leaving the exit pupil does not come to a point, but results in these two curved lines; the term 'astigmatism' implies this pencil or cone without a point.

The 'image' *hg* is termed the *primary*, or *tangential*, *image* and extends in the direction at right angles to the plane containing the axis *OQ* and the object (i.e. in the *y*-direction). The 'image' *fe* is called the *secondary*, or *sagittal*, *image* and extends in the *z*-direction, parallel to the length of the object. These images are illustrated in Fig. 2.12. The separation along *OQ* between the tangential and sagittal images is the *astigmatic difference*.

Between *hg* and *fe*, the image is in general elliptical. The cross-section of the astigmatic pencil of rays is least midway between *hg* and *fe*; this is the position of best focus of a point object, (the image then being a circle, called the *circle of least confusion*). If an object in the form of a spoked wheel in a plane parallel to *yOz* were used with an optical system having an exit pupil as shown in Fig. 2.11, the image formed would show the rim of the wheel sharply focused in the region *hg* (hence the term 'tangential'), whilst the spokes would be sharply focused in the region *fe* (the term 'sagittal' comes from the Latin word for an arrow).

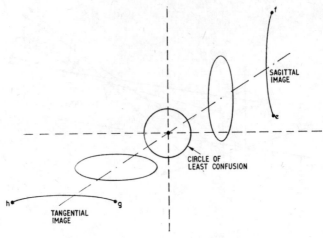

Fig. 2.12. Nature of the image formed when astigmatism is present

2.12 Curvature of the Field

This aberration, important when an extended object is used, is closely associated with astigmatism. Consider an object which extends laterally and symmetrically about the optical axis of a system. In the plane of the diagram (Fig. 2.13), the object will be a line and the images will be in the form of parabolae in this

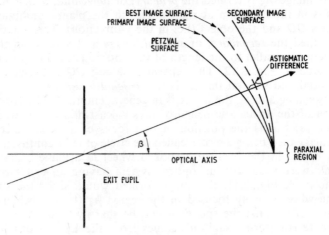

Fig. 2.13. Curvature of the field

plane. In three dimensions, the object will be a plane and the images will be paraboloidal surfaces. The primary, or tangential, surface will be separated by the astigmatic difference from the secondary, or sagittal, surface. This astigmatic difference increases with the angle β formed by a ray from a point on the line object which passes through the centre of the exit pupil of the optical system.

If by some means the system is corrected for astigmatism, the two surfaces, primary and secondary, will combine to form a single surface, also parabolic, called the *Petzval surface*. This will generally have more curvature than either the primary or secondary surface. On the other hand, if the surface of the best image is made flat, astigmatism will be generally greater towards the edges of the field. Astigmatism and curvature of the field may both be reduced by appropriate positioning of stops in the optical system, and by shaping of lens surfaces. For two separated lenses, reduction of these two aberrations is achieved by stop positioning and satisfying the *Petzval condition*:

$$n_1 f_1 + n_2 f_2 = 0$$

where n_1 and n_2 are the refractive indices of the materials of the lenses relative to the medium in which they are situated (normally air for which $n = 1$), and f_1 and f_2 are the focal lengths of the respective lenses.

2.13 Distortion

In general, for an object which extends symmetrically and laterally about the optical axis, the outer parts which are further from the optical axis will be magnified differently from the central parts near the optical axis. This leads to an arrangement of image points geometrically different from that of corresponding object points. This aberration is known as distortion.

As an example, consider the effect of a stop S placed in front of a converging lens [Fig. 2.14(a)]. If the lens suffers from spherical aberration, a cone of rays from an non-axial point P_2 on the object will be magnified more than a cone from the paraxial point P_1 on the object. Hence, the outer parts of the object will be more magnified than the inner parts, and an image of a square object in a plane perpendicular to the optical axis will be as shown in Fig. 2.14(b). This type of distortion is called *pincushion distortion*. However, if the stop S is placed after a converging lens L (on the opposite side from the object), the magnification is greater for the paraxial central parts of the object

76

UNIVERSITY OPTICS

than for the outer parts, so a barrel-shaped image of a square plane object is obtained; this defect is known as *barrel distortion* [Fig. 2.14(c)].

Thus, pin-cushion distortion results when an optical system has other aberrations whereby the central portions of the image

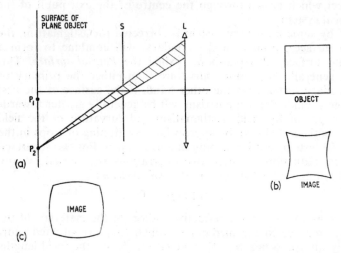

(a)

(b)

(c)

Fig. 2.14. Distortion: (a) effect of a stop placed in front of a converging lens; (b) pin-cushion distortion; (c) barrel distortion

are less magnified than the outside portions; on the other hand, barrel distortion occurs when the central portions are the more magnified.

To understand how distortion may be minimised, consider the optical system represented by Fig. 2.15, in which E is the centre of the entrance pupil and E' is the centre of the exit pupil. Let P be a paraxial point in the plane object at a distance h from the optical axis where $\angle AEP = \omega$, and let Q be the conjugate paraxial point in the image plane at a distance h' from the optical axis where $\angle QE'B = \omega'$. For a point P_1 in the outer part of the object plane at a distance h_1 from the optical axis, let $\angle AEP_1 = \omega_1$ and let the corresponding image Q_1 be at a separation h'_1 from the optical axis where $\angle Q_1E'B = \omega'_1$.

If it is assumed that the entrance and exit pupils of the optical system are corrected for spherical aberration for rays from the object points concerned,

$$\frac{h}{h_1} = \frac{\tan \omega}{\tan \omega_1}$$

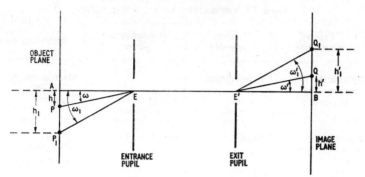

Fig. 2.15. Reduction of distortion

and

$$\frac{h'}{h'_1} = \frac{\tan \omega'}{\tan \omega'_1}$$

If the magnification is to be the same for central and outer regions of the image, i.e. if distortion is to be zero,

$$\frac{h'_1}{h_1} = \frac{h'}{h}$$

Therefore,

$$\frac{\tan \omega'_1}{\tan \omega_1} = \frac{\tan \omega'}{\tan \omega}$$

i.e.

$$\frac{\tan \omega_i}{\tan \omega_o} = \text{constant}$$

where ω_o is the angle for any point in the object and ω_i the angle for the corresponding image point.

This result is known as the *tangent condition*. It is not of wide application because it is true only if the pupils E and E' have no aberration. In practice, correction for spherical aberration is for the object and image and not for the entrance pupil and the exit pupil.

2.14 Summary Showing the Dependence of the Seidel Aberrations on Aperture and Field of View

Table 2.1 shows the manner in which the various aberrations in a coaxial system depend upon the radius ϱ of the aperture of the

Table 2.1. ABERRATIONS OF A COAXIAL SYSTEM

Type of aberration	Cones affected	Dependence on ϱ and y	Instruments
Spherical: longitudinal λ transverse τ	Symmetrical	$\lambda \propto \varrho^2$ $\tau \propto \varrho^3$	Microscope, telescope, camera
Coma	Skew	$\tan \beta^* \propto \varrho^2 y$	Camera, telescope, microscope
Astigmatism and curvature of field	Skew	Astigmatic difference $\propto \varrho y^2$	Camera, telescope, microscope
Distortion	Symmetrical and skew	Magnifying power $\propto y^3$	Camera, microscope, telescope

* The angle β concerned is shown in **Fig. 2.8(b)**.

system and upon the perpendicular distance y between a point in an object plane and the optical axis (this distance being related to the field of view). Within the column headed 'cones affected', a symmetrical cone of rays is one with its apex on the optical axis, whereas a skew cone is one which originates from a non-axial point. Also listed are the chief optical instruments affected, in the appropriate order for each aberration.

2.15 Chromatic Aberration

The refractive index of a dispersive optical medium is a function of the wavelength of the light traversing the medium. Since the power of an optical system depends on the values of the refractive index, it will also vary with wavelength. This variation in power will give rise to two effects.

2.15.1 LONGITUDINAL CHROMATIC ABERRATION

An incident beam of heterogenous light (i.e. light containing components of various wavelengths, white light being the most usual in practice) which is parallel to the optical axis of the system will not be converged (or diverged) to a given focal point after passage through the system. Instead, there will be various focal points for the different wavelengths: for example, second foci at F'_{λ_1} for light of wavelength λ_1 and F'_{λ_2} for light of wavelength λ_2. These foci will be on the optical axis.

The longitudinal chromatic aberration [Fig. 2.16(a)] is usually specified as the axial separation between the focal point F'_C and

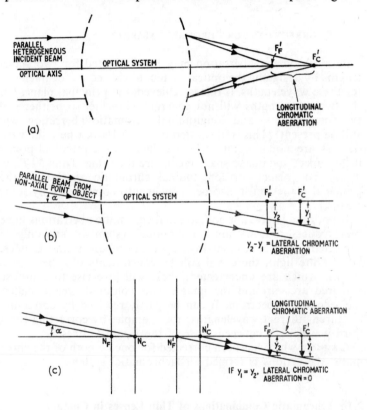

Fig. 2.16. Chromatic aberration

the focal point F'_F. Here, F'_C is the focal point for light of wavelength 6,563 Å (1 Å = 1 Ångstrom unit = 10^{-8} cm) which is that of the red C line in the hydrogen spectrum, and F'_F is the focal point for the green-blue F hydrogen line of wavelength 4,861 Å. These two reference wavelengths are those of the Fraunhofer C line and the Fraunhofer F line.

2.15.2 LATERAL CHROMATIC ABERRATION

The lateral magnification of the optical system will also vary with wavelength giving rise to lateral, or transverse, chromatic aberration. This may be measured by the difference between the lateral displacements from the optical axis of the images formed of a distant non-axial point object for the Fraunhofer F and C wavelengths [Fig. 2.16(b)].

2.15.3 CORRECTIONS FOR CHROMATIC ABERRATION

Lateral chromatic aberration may be eliminated for two wavelengths by designing the optical system to give equal focal lengths for these wavelengths. If this is achieved, the principal planes for the two wavelengths will not generally coincide, so neither will the focal points, and longitudinal chromatic aberration will still be present. [This is illustrated in Fig. 2.16(c), where the various Ns are nodal points, which coincide with principal points if the object and image space media are the same. Thus $N_F' F_F' = N_C' F_C'$. To correct for longitudinal chromatic aberration, the principal planes must also be brought into coincidence by some means.]

If both the lateral and longitudinal chromatic aberrations have been reduced to a minimum, the optical system will still only be corrected for two wavelengths, say those of the F and C lines. With white light, there will still be constituents of other wavelengths which are uncorrected; these will give rise to a diffuse coloured area around the image. This coloured area is called the secondary spectrum. It can be partly reduced by correcting for three different wavelengths; for example, by correcting for a third wavelength between those of the F and C lines.

A system which has been corrected for one or both of the chromatic aberrations is termed an 'achromatic' system.

2.16 Achromatic Combinations of Thin Lenses in Contact

An optical system comprises a number N of thin lenses 1, 2, 3, ... p, ..., N which are coaxial and succeed one another in contact in air. Let the pth element of this assembly have successive surfaces of radii $_p r_1$ and $_p r_2$, be made of material of refractive index n_p for a given wavelength, and have an optical power F_p and focal length f_p. Now, $F_p = 1/f_p$ and, in Equation 1.10, $n = n' = 1$ for air, $n_m = n_p$, $r' = {}_p r_2$ and $r = {}_p r_1$.

Therefore,

$$F_p = \frac{1}{f_p} = (n_p-1)\left(\frac{1}{{}_pr_1} - \frac{1}{{}_pr_2}\right) = (n_p-1)C_p \qquad (2.10)$$

where C_p is a constant for the pth element independent of wavelength.

If the wavelength λ of the light passing through the optical system changes by $\delta\lambda$, F_p changes by δF_p where, from Equation 2.10,

$$\delta F_p = C_p\,\delta n_p$$

and

$$\frac{\delta F_p}{F_p} = \frac{\delta n_p}{n_p - 1} \qquad (2.11)$$

If the optical power of the whole system of N lenses is F for a wavelength λ,

$$F = \sum_{p=1}^{p=N} F_p$$

since the power of a number of lenses in contact is the sum of the individual powers. Therefore, for a change in λ of $\delta\lambda$,

$$\delta F = \sum_{p=1}^{p=N} \delta F_p = \sum_{p=1}^{p=N} \frac{F_p\,\delta n_p}{n_p-1}$$

from Equation 2.11.

If the combination of lenses in contact is to be achromatic,

$$\delta F = \sum_{p=1}^{p=N} \omega_p F_p = 0 \qquad (2.12)$$

where

$$\omega_p = \frac{\delta n_p}{n_p - 1} \qquad (2.13)$$

and is defined as the *dispersive power* of the material of the pth element for the wavelengths λ and increment $\delta\lambda$.

Usually, δn_p is taken as the difference between the refractive index of the material at the wavelength of the Fraunhofer F line and this index for the Fraunhofer C line; n_p, the value about which variation with wavelength occurs, is conventionally taken for the orange-yellow D line in the sodium spectrum. (There are two D lines, of wavelengths 5,890 Å and 5,896 Å, the mean value 5,893 Å being used). The dispersive power ω of a refractive medium is then given by:

$$\omega = \frac{n_F - n_C}{n_D - 1} \qquad (2.14)$$

The *reciprocal dispersive power* $V = 1/\omega$ is also known as the *constringence*.

In the special case of two thin lenses in contact in air, i.e. $N = 2$, an *achromatic doublet* is concerned, for which Equation 2.12 reduces to:

$$\delta F = \sum_{p=1}^{p=2} \omega_p F_p = \omega_1 F_1 + \omega_2 F_2 = \frac{\omega_1}{f_1} + \frac{\omega_2}{f_2} = 0 \qquad (2.15)$$

In a typical example where the doublet is to be converging and made from crown and flint glass components, Equation 2.15 becomes:

$$\frac{F_{crown}}{F_{flint}} = \frac{f_{flint}}{f_{crown}} = -\frac{\omega_{flint}}{\omega_{crown}} \qquad (2.16)$$

A converging doublet must clearly comprise a converging lens and a diverging lens in contact because the powers must be respectively positive and negative to satisfy Equation 2.16. Further, as $\omega_{flint} > \omega_{crown}$, F_{crown} must be greater than F_{flint}, the converging element must be of crown glass and the diverging one of flint glass.

It is possible to design an achromatic doublet which is also free of spherical aberration. To do this, use is made of the Lister doublet (Section 2.4) comprising one converging and one diverging thin lens; the ratio f_1/f_2 of these lenses will generally be a function of the zone radius and the shape and position factors. The calculated value of f_1/f_2 can be made also to satisfy Equation 2.15; thus, an achromatic doublet, free from spherical aberration can be achieved.

If an achromatic doublet is made so that it also satisfies the Petzval condition (Section 2.12), it will be free of field curvature (i.e. it will give a flat field). The equations to be simultaneously satisfied are now, from Section 2.12,

$$n_1 f_1 = -n_2 f_2$$

and, from Equation 2.15,

$$\frac{f_1}{f_2} = -\frac{\omega_1}{\omega_2}$$

Hence, the condition is that, numerically,

$$\frac{\omega_1}{\omega_2} = \frac{n_2}{n_1}$$

This demands that the material of greater refractive index (which is used for one of the lenses) must have the smaller dispersive power. Most glasses will not satisfy this condition, but there are certain barium glasses which can be used to design an *apochromatic* objective lens with small chromatic aberration and a substantially flat field.

2.17 Achromatic Combinations of Separated Thin Lenses

Two lenses made of the same material, of focal lengths f_1 and f_2, and separated by a distance d in air will, as a combination, have an optical power F given from Equation 1.20 by

$$F = F_1 + F_2 - dF_1F_2$$

since $_1n_2 = 1$ for air, $F_1 = 1/f_1$ and $F_2 = 1/f_2$. This equation will hold for a given wavelength λ. If λ is changed by $\delta\lambda$,

$$\delta F = \delta F_1 + \delta F_2 - d(F_2\delta F_1 + F_1\delta F_2)$$

If ω is the dispersive power of the material of these lenses, $\delta F_1 = \omega F_1$ and $\delta F_2 = \omega F_2$ from Equation 2.11. The combination will be achromatic, therefore, if

$$\delta F = \omega F_1 + \omega F_2 - 2\omega dF_1F_2 = 0$$

i.e. if

$$F_1 + F_2 - 2dF_1F_2 = 0$$

or

$$d = \frac{F_1 + F_2}{2F_1F_2} = \frac{f_1 + f_2}{2} \qquad (2.17)$$

As ω does not appear in Equation 2.17, this combination of two thin lenses separated by a specific distance d is corrected for *lateral* chromatic aberration for all wavelengths because the focal length and the optical power is independent of wavelength. However, the principal planes do not coincide for the various wavelengths, so longitudinal chromatic aberration remains. This aberration can only be minimised for two wavelengths by forming such a combination from two achromatic doublets or triplets.

The pair of thin single lenses separated by the distance d given by Equation 2.17 can also be designed to give minimum spherical aberration or to satisfy the Petzval condition (Section 2.12), as well as to give minimum lateral chromatic aberration.

Exercise 2

1. Define the main stops and pupils of an optical system and describe the function of stops in minimising aberrations.

A thin lens of diameter 5 cm and focal length 4 cm has a 3 cm diameter stop placed 2 cm in front of it. Calculate the position and size of the exit pupil, and draw in a diagram roughly to scale the marginal and chief rays from a point 8 cm in front of the lens and 1·5 cm off the axis to the corresponding image point.   (L. P.)

2. Describe the primary monochromatic aberrations of a simple thin lens, indicating how they depend upon the aperture and field of view.   (L. P.)

3. Discuss the Abbe sine condition in relation to the aberration of lenses. A lens system composed of two thin lenses, one converging and the other diverging, each of focal length 10 cm, with their centres 5 cm apart, forms an image of a small luminous coaxially centred object on a screen. If the image is the same size as the object what is the separation of the object and the screen? Assuming the thin lens formula, prove any other formula used.   (L. P.)

4. Establish the condition that two thin lenses in contact may form an achromatic combination.
A converging achromatic cemented doublet of focal length 50 cm is to be made from glasses A and B, where the refractive index n_D for glass A is 1·516 and $n_F - n_C = 0.0081$; for glass B, n_D is 1·648 and $n_F - n_C = 0.0190$. If the diverging lens of the combination is to have one face plane, calculate the radii of curvature of the faces of the converging lens required.

5. A thick lens is made by grinding both ends of a glass cylinder of length l and refractive index n to convex surfaces with radius of curvature r. Find the positions of the unit (principal) points, and the principal foci of this system, defining carefully what is meant by them.
Show that if $r = 0.5\, l\, (1 - 1/n^2)$ the focal length is almost independent of small changes of n. In what sense can the lens be said to be achromatic in this case?   (L. Anc.)

6. What are the cardinal points of an optical system? Define the terms used.
(a) Two similar plano-convex thin lenses are made of glass of refractive index 1·5 for sodium light and their convex surfaces have a radius of 2·5 cm. Their plane surfaces are towards each other and are separated by 5·0 cm. Calculate the focal length of the system and locate its focal points.
(b) Make similar calculations for the case in which the whole of the space of 5·0 cm between the plane faces is filled with glass of the same refractive index.
Explain in what sense the system (a) may be said to be achromatic. (L. G.)

7. Write an essay on the monochromatic aberrations of optical systems.
                                    (L. G.)

8. What is spherical aberration?
A plano-convex lens is made of glass of refractive index 1·5, and has a

radius of curvature of 20 cm, a thickness at the centre of 3 cm, and a diameter of 10 cm. Find the variation in focal length when (a) the curved surface, (b) the plane surface, faces the incident light. Discuss the practical significance of the result.   (G. I. P.)

9. Derive the condition for a pair of thin lenses to be an achromatic combination.

Two sorts of glass with the properties given below are to be used to make a cemented achromat.

(i) $n = 1\frac{2}{3}$ $\omega = 0\cdot0244$

(ii) $n = 1\frac{1}{2}$ $\omega = 0\cdot0183$

The combination is to have a focal length of 24 cm. The radius of curvature of the external surface of the positive component is to be 12 cm.

Find the radii of curvature of the other lens surfaces and show that the lens will have little spherical aberration also for object and image of equal size.   (Aston)

CHAPTER 3

Optical Instruments

3.1 Introduction

The optical instruments of concern in this chapter are the human eye, the telescope, the microscope and the camera. The treatment is geometrical; full consideration is not taken of the resolving power, which is intimately linked with the instrument aperture and the wavelength of the light involved. It must be realised that, even if a system is free from all the aberrations described in Chapter 2, its performance will be limited by the diffraction pattern formed by the exit pupil aperture. Considerations of resolving power are deferred until Chapter 12, so that the subject can be studied after the appropriate wave theory.

There is, furthermore, the question of the photometric qualities of the image (for example, its brightness). This is considered in photometry in Chapter 6.

3.2 The Human Eye

Since in many observations with optical instruments the human eye forms a part of the optical system, it clearly merits study. It has, furthermore, a structure which is undoubtedly much more remarkable than that of any man-made optical instrument. However, the concern here is not the intimate detail of its fascinating anatomical structure but its physical functioning, particularly in relation to visual observation aided by an optical instrument.

A horizontal section through a human right eye is shown in Fig. 3.1. The whole structure—the *eyeball*—is located in a cavity in the human skull called the *orbit*. The eyeball is roughly spherical except that the front or anterior portion (nearest the object viewed), called the *cornea*, has a shorter radius of curvature (and so bulges more convexly) than the remainder and extends over about one-sixth of the eyeball. The cornea is of a hard transparent material, whereas the remaining five-sixths of the eyeball surface—the *sclerotic* coat—is opaque.

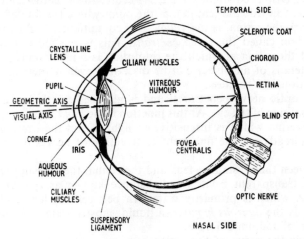

Fig. 3.1. Horizontal section of the human (right) eye

The coloured circular *iris* is a membrane attached where the sclerotic merges into the cornea. Slightly off-centre and displaced towards the nose, the iris contains a circular aperture called the *pupil*, which has a maximum diameter of about 8 mm. This diameter diminishes involuntarily as the iris contracts when the incident light intensity increases. Thus, the pupil will be wide open in a dark room and diminish to about 2 mm in diameter in sunlight. The amount of light entering the *crystalline lens* immediately behind the pupil is thus regulated. Optically speaking, this crystalline lens is not simple but is a complex system made in many layers of tough transparent tissue of refractive index varying from about 1·37 in the outer parts to about 1·42 at the centre. The annular *suspensory ligament* supports the lens from the *ciliary muscles*, which form the ring-shaped ciliary body. These muscles are able to adjust the shape and thickness of the crystalline lens to form a focused image on the *retina*. This adjustment of shape enables the focal length of the crystalline lens to be varied, so that a sharp image is obtainable irrespective of whether the distance to the object is between 25 cm (or less) and infinity; thus, the eye is given the power of *accommodation*.

The retina over the interior of the posterior surface of the eye is a highly complex network of light-sensitive nerves, which connect via the *optic nerve* to the brain. Immediately within the sclerotic covering the eye is the *choroid*, the interior of which contains black pigment cells which absorb superfluous light entering the eye. Displaced slightly from the centre towards the temporal side (i.e. towards the temple), at the point where the line of sight intersects the retina, is the *yellow spot* containing a slight central depression called the *fovea centralis*. This is the most sensitive part of the retina at which the resolution is a maximum: at this part, details of an object can be discerned even if their angular separation is 1' or less.

The optic nerve penetrates, and is attached to, the retina on the nasal side of the eye. At this junction is the *blind spot*—a region of the retina which is insensitive to light. From the optic nerve there spreads out the fine weblike structure of nerves of the retina.

Between the crystalline lens and the retina, the eyeball is filled with a transparent gelatinous substance known as the *vitreous humour*, which is primarily water. In between the lens and the cornea is the *aqueous humour*, a liquid, mostly water containing a trace of sodium chloride.

Within the tissue of the retina and its complex nerve system is the *bacillary layer* of *rods* and *cones*, of which the former are the more numerous. These rods and cones are the light-sensitive cells: they contain photochemical substances which undergo changes on receiving light. The rods are able to detect very small amounts of light but cannot readily distinguish colour—only black, blue, greys and white. In a dim light, as at night, it is the rods which are primarily responsible for vision; night vision is with poor resolution of detail and little or no perception of colour. The main instruments of day vision are the cones, which are congregated mostly in a small area of about 1·5 mm in diameter on the temporal side about the fovea centralis. The central region of this area, over a circle of about 0·25 mm in diameter— the fovea—is the region of most distinct vision and contains only cones; rods are absent. It is sensitive to the incidence of as little as 30 photons per second of green light. Outwards from the fovea, the number of cones per unit area diminishes to becomes near zero at the periphery of the retina, whilst the number of rods increases.

The rods contain a pink colouring matter, known as *visual purple*, which is very sensitive to light. In a dark room or poor light, visual purple is secreted slowly to render the rods more sensitive. Thus, in a very dim light or at night, objects not perceived at first become visible after some minutes as visual purple is secreted, the maximum sensitivity being after about 40 min.

Under normal conditions of illumination, the muscles controlling the eyeball, which is rotatable about three axes (motion up and down, to left and right, and, to a smaller extent, inwards in squinting are possible), and the ciliary muscles will adjust to form a sharp image in the foveal region. This observation is by a scanning process but for simplicity the image is assumed to be stationary in calculations involving the eye.

When a normal eye is at rest (i.e. when there is no exertion of the ciliary muscles), it is focused at infinity. The power of adjusting the crystalline lens of a normal eye is limited so that objects closer than about 25 cm cannot be seen clearly, though this distance is considerably less for young people. This distance of 25 cm (10 in) is called the *near distance* of the eye. Owing to defects in the crystalline lens or in the shape of the eyeball, the short-sighted eye will have a near distance less than 25 cm, and the long-sighted eye will have a larger value. The normal eye will clearly focus images of distant objects but, for the short-sighted eye, accommodation will not begin until the object is at a point known as the *far point*. On the other hand, for the long-sighted eye, some accommodation will occur when viewing distant objects

Gullstrand, on the basis of a great number of observations of the refractive indices and radii of curvature involved in human eyes, has suggested an average eye having the properties given in Table 3.1, which is related to Fig. 3.2. The positions of the

Table 3.1. PROPERTIES OF AN AVERAGE EYE

Name of point or distance	Distance from the anterior pole A of the cornea (positive towards the retina) (mm)
First principal point P	1·348
Second principal point P'	1·602
First focal point F	−15·707
Second focal point F'	24·387
First focal length f	−17·055
Second focal length f'	22·785
Fovea centralis M	24·000

The fovea is depressed about 0·387 mm.

nodal points may be found in the usual way. For approximate calculations, the principal points P and P' may be regarded as coinciding at a distance of 1·5 mm from the anterior pole A of the cornea.

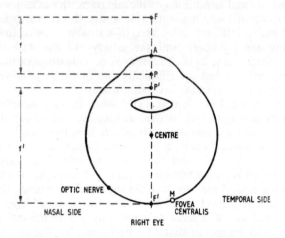

Fig. 3.2. Equivalent optical system of the human eye

Emsley has shown that the above system is very nearly equivalent to a single spherical refracting surface of radius 5·55 mm of power 60 dioptres (focal length $= \frac{1}{60}$ m), containing a medium of refractive index 1·333 relative to air as unity, and with its pole displaced 2 mm towards the retina from the anterior pole of the cornea.

3.3 The Telescope

The refracting telescope comprises a converging objective lens system of long focal length coaxial with a separate shorter focal length eyepiece lens system which may be converging or diverging. In the 'normal' adjustment of a telescope, the second focal point of the objective system coincides with the first focal point of the eyepiece. Parallel light rays entering the objective from a distant object are converged to the common focal points and leave the eyepiece as parallel rays to a final image at infinity.

The *astronomical telescope* has a converging eyepiece, the *Galilean telescope* has a diverging eyepiece. Simple models will have a single convex (converging) lens as the objective and a simple convex or concave (diverging) lens as the eyepiece.

When the objective and eyepiece are arranged in the mounting tube, the entrance aperture is usually the objective itself, which therefore also forms the entrance pupil (Section 1.13).

In the astronomical telescope (Fig. 3.3), let P_o and P_o' be the first and second principal points of the objective with corresponding principal planes through them perpendicular to the optical axis, whilst P_E and P_E' are the corresponding points for the eyepiece. The focal points are likewise denoted by F_o and F_o' for the objective, and F_E and F_E' for the eyepiece; the focal lengths are f_o and f_E respectively, where the first and second focal lengths for each lens are the same in the usual event of the medium external to the lenses being air (when the principal and the corresponding nodal points also coincide). F_o' and F_E will coincide in normal adjustment.

The *angular magnification* m_A is defined as the angle subtended at the observer's eye (strictly at the first nodal point of the eye) by the image divided by the angle subtended by the object. For a telescope in normal adjustment, the apparent image and the object will both be very distant. In Fig. 3.3, consider a very distant object of height h above the optical axis of the telescope; the point on this object furthest from the optical axis produces parallel rays at an angle α to the optical axis. Rays meeting the

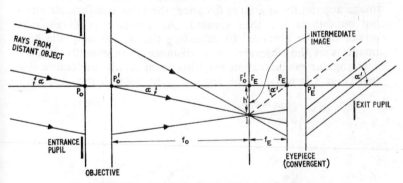

Fig. 3.3. The astronomical telescope in normal adjustment

first principal plane through P_o, which here is also the first nodal plane, will leave the second principal plane through P_o' so that the angle of the central ray is still at angle α to the optical axis. Three rays, after refraction by the objective, will intersect in the focal plane through F_o', which coincides with F_E. The intermediate image formed will be of height h'. Leaving these common focal planes, the rays will be refracted by the eyepiece to emerge in a parallel beam at an angle α' to the optical axis and to enter the eye.

In the diagram, a ray IP_E from the intermediate image to the first principal (nodal) plane P_E will emerge from P_E' at α' to the optical axis. $\angle F_E P_E I$ is obviously also α'. The angular magnification $m_A = \alpha'/\alpha$. As these angles are small when the object is very distant, $\tan \alpha = \alpha$ and $\tan \alpha = \alpha'$. Therefore,

$$m_A = \frac{\alpha'}{\alpha} = \frac{h'}{h} = \frac{f_o}{f_E} \qquad (3.1)$$

The aperture stop of the objective forms the entrance pupil. The exit pupil, which is the image of the aperture stop in all the optical elements following it, should have a diameter less than the diameter of the entrance pupil of the eye under normal conditions of illumination *(photopic vision);* otherwise, the iris of the eye will form the aperture stop of the system and a loss of transmitted light will result. The eye should be placed so that the exit pupil of the telescope is coincident with the entrance pupil of the eye. To aid the observer in correct location of the eye, a ring called the *eye-ring* is mounted at an appropriate position on the outlet of the eyepiece.

In the normal adjustment of the astronomical telescope, the crystalline lens is relaxed for the normal eye as it is focused on an image apparently at a large distance; the most comfortable viewing conditions are thus created. An increase in magnifying power may be obtained by adjusting the system until the final image is at the observer's near distance D (nominally 25 cm). The angular magnification resulting from an all-converging system (Fig. 3.4) will now be angle α' subtended at the observer's

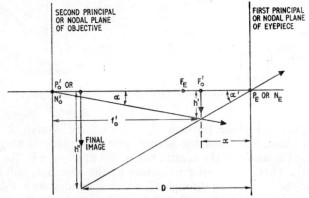

Fig. 3.4. *Astronomical telescope adjusted so that the final image is at the least distance of distinct vision*

eye by the final image divided by the angle subtended at the un-aided eye by the distant object. Note that F_o' and F_E do not now coincide; the separation between the objective and the eyepiece has been altered by appropriate adjustment of the telescope.

With reference to Fig. 3.4, the angular magnification m_A is given by

$$m_A = \frac{\alpha'}{\alpha} = \frac{h''/D}{h'/f_o'} = \frac{h''f_o'}{h'D} \tag{3.2}$$

providing that α' and α are small angles. For the eyepiece system:

$$\frac{1}{f_E'} = \frac{1}{x} - \frac{1}{D}$$

(Note the incident vergence is positive and the emergent ver-gence is negative, as explained in Section 1.5.) Since $f_E = f_E'$ for the eyepiece in air,

$$\frac{D}{f_E} = \frac{D}{x} - 1 = \frac{h''}{h'} - 1 \tag{3.3}$$

Substitution from Equation 3.3 into Equation 3.2 gives:

$$m_A = \left(\frac{D}{f_E} + 1\right)\frac{f_o'}{D} = \frac{f_o'}{f_E} + \frac{f_o'}{D} \tag{3.4}$$

($f_o' = f_o$ for the objective in air). Thus, this adjustment gives an advantage of f_o/D compared with normal adjustment (Equa-tion 3.1).

The Galilean telescope has a converging objective but a diverg-ing eyepiece (Fig. 3.5). The parallel rays from the distant object are brought to a focus in the second focal plane through the sec-ond focal point F_o' of the objective, but this focal plane is now

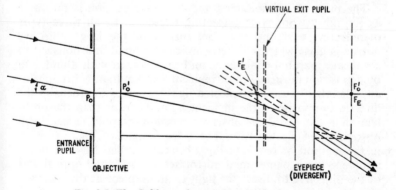

Fig. 3.5. The Galilean telescope in normal adjustment

beyond the eyepiece. F_o' and F_E, the first focal point of the eyepiece, coincide but the rays entering the eye are from a virtual image. As the exit pupil is the image of the aperture stop (usually the objective) in the eyepiece, this pupil is also virtual. Equations for magnification of the Galilean telescope are obtained as before.

In the astronomical telescope with a converging eyepiece, the field stop (Section 1.13) coincides with the second focal plane of the objective and so with the first focal plane of the eyepiece, where the intermediate or primary image is formed. In the Galilean telescope, there is no real intermediate image and so no defined field stop. The field of view is then usually limited by the objective if the eye is immediately behind the eyepiece, and is more restricted than in a comparable instrument with a converging eyepiece.

In the astronomical telescope, the image is real and inverted; in the Galilean instrument, the image is virtual and upright.

The *reflecting telescope* has a converging mirror as the objective instead of a lens. Three arrangements are used: the Newtonian [Fig. 3.6(a)], the Cassegrainian [Fig. 3.6(b)] and the Gregorian [Fig. 3.6(c)]. The latter two arrangements are favoured. The resolving power of a telescope is directly proportional to the diameter of its objective, as can be shown by diffraction theory (Section 12.3). Large astronomical telescopes for studying very distant constellations (such as the 100 in objective diameter instrument at Mount Wilson and the 200 in one at Mount Palomar, both in California, USA) have mirror objectives because of the difficulty of providing high-quality optical glass for the fabrication of lenses of such size and with low light absorption; moreover, the use of vacuum-deposited aluminium on the front surface mirror enables photography to be undertaken in ultraviolet light which would not be transmitted through glasses.

The principle of obtaining angular magnification is the same as for the astronomical refracting telescope. In all three types of reflecting telescope, the front surface of the large objective mirror is glass which has been optically ground and polished to a concave paraboloidal shape and then coated with aluminium by deposition *in vacuo;* thus, incident parallel light is brought to an aplanatic focus. The reflected light travels to the focus in the direction opposite to the incident light. There is then the problem that the eyepiece, observer or photographic camera will obstruct the reflected rays. The three designs of Fig. 3.6 differ in the method of reducing this obstruction. The Newtonian instrument has a plane small mirror set at 45° to the optical axis of the objective to direct the light to an eyepiece at the side; the Cassegrainian arrangement utilises a small convex hyperboloid-

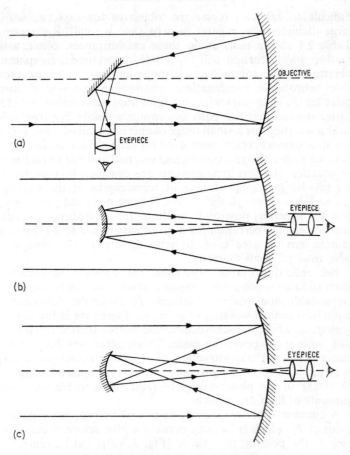

(a)

(b)

(c)

Fig. 3.6. Reflecting telescopes: (a) Newtonian;
(b) Cassegrainian; (c) Gregorian

al mirror to return the light to an eyepiece behind a small central hole in the objective; the Gregory telescope is similar to the Cassegrainian one except that a concave ellipsoidal mirror is used.

3.4 Telescope Objectives

The focal length of a telescope objective is large compared with that of the eyepiece so that high magnification can be obtained. Values up to 10 m and more are not uncommon. As it is very difficult to fabricate corrected objective lens systems with a large diameter, the angular field of view is small. Reference to Table 2.1 shows that, under these circumstances, coma, astigmatism and distortion will be of less importance than spherical aberration. The commonest telescope lens objectives are therefore achromatic combinations comprising converging crown glass lenses in contact with diverging flint glass lenses. In 1830, Lister showed that two pairs of conjugate points free from spherical aberration for a small range of apertures outside and including the paraxial region must exist for such a combination. These pairs comprise two real points, and one real and one virtual point. In practice, the best arrangements are designed by repeated trials involving tracing of rays of wavelengths of the hydrogen C and F lines through the system. If cemented doublets are used, it is not generally possible to minimise spherical aberration, coma and chromatic aberration for a given zone range so a three-component lens must be used. In general, this is only possible for objectives of small diameter.

For reflecting mirror objectives, the question of chromatic aberration does not arise. Spherical aberration can be minimised by suitable non-spherical surfaces. To avoid the difficulties of aspherical optical working of glass, an alternative is the *Schmidt system*, in which a concave spherical mirror is used in conjunction with a compensating plate. This system has been particularly useful in the construction of reflecting telescopes of large aperture (and so large light-gathering power); such telescopes are necessary in the photography of faint stars, where very small amounts of light are received.

A concave spherical mirror, as is well known, has a paraxial focus at F_p which is halfway between C, the centre of curvature, and A, the pole of the mirror [Fig. 3.7(a)]; but marginal rays are brought to a focus nearer A at a point F_M, the point of intersection of incident parallel rays passing through the marginal zone of the aperture stop. The separation F_pF_M is a measure of the longitudinal spherical aberration, and there is corresponding transverse spherical aberration. To ensure that the incident ray OM parallel to CA in Fig. 3.7(a) would pass through the paraxial focus F_p after reflection, its path would have to be modified to $O'M$, where $\angle O'MC = \angle F_pMC$, to satisfy the reflection laws at M. Thus, the parallel ray to a marginal zone in Fig. 3.7(b) needs to be diverged outwards—which is achieved by the corrector plate—before incidence on the concave spherical mirror.

The corrector plate placed perpendicularly across the incident parallel rays could therefore be like a shallow concave lens, the path retardation introduced in this plate being proportional to ϱ^4, where ϱ is the radius of any zone. However, such a plate, in providing the necessary slight divergence of the incident beam, would introduce chromatic aberration because of dispersion. To reduce this chromatic aberration to a minimum, the maximum ray divergence must be as small as possible. The Schmidt

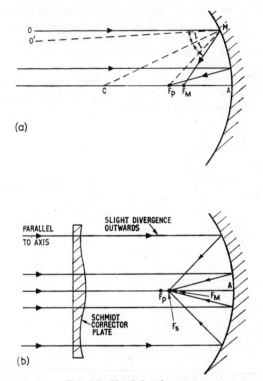

Fig. 3.7. The Schmidt system

corrector plate as used, therefore, is a compromise to give minimum chromatic and spherical aberrations. This is arranged by causing the incident paraxial and marginal rays to pass, after reflection, through a circle of least confusion which is as small as possible and which is at F_s between F_p and F_M such that the distance $F_M F_s = \frac{1}{4} F_M F_p$. The corrector plate profile as represented in Fig. 3.7(b) achieves this: it has an optical path length through its central and marginal zones which is greater than that through intermediate zones.

3.5 The Microscope

The microscope is similar to the astronomical refracting tele-scope in that it comprises a combination of two coaxial separated converging systems but with the difference that the objective lens system has a shorter (usually much shorter) focal length than the eyepiece system. The objective forms a real inverted image of a close object; this intermediate image is then observ-ed by the eyepiece as a magnified, virtual and inverted image. This instrument is called the compound microscope to distin-guish it from the simple microscope which is a single converging lens or hand magnifier.

The objective–eyepiece separation is adjusted to give the great-est magnification when the final image is formed at the least distance of distinct vision, D (nominally 25 cm).

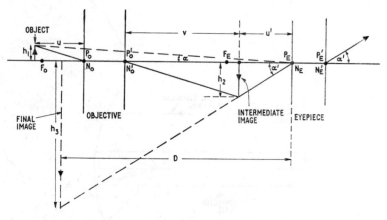

Fig. 3.8. Principle of the compound microscope

In Fig. 3.8, which applies to the usual situation where the ob-jective and eyepiece are in air, P_o and P'_o are the first and second principal points respectively of the objective lens system and these coincide with the corresponding nodal points N_o and N'_o. At a considerable distance (compared with focal lengths of the lenses) away from the objective is the coaxial eyepiece with first and sec-ond principal points at P_E and P'_E respectively; again, these coincide with nodal points N_E and N'_E respectively.

An object of height h_1, with its base on the optical axis just before the first principal focus F_o of the objective and at a distance u from P_o, will give rise to a real, inverted intermediate image of height h_2 at an axial distance v from P'_o. This intermediate image, at an axial distance u' from P_E', is arranged to be situated between P_E and F_E, the first principal focus of the eyepiece, so that it gives rise to a final virtual image of height h_3 which, with appropriate adjustment of the microscope, is at an axial distance D from P_E.

The angular magnification m_A of the instrument is defined as the angle α' subtended at the observer's eye by the image at the near distance D divided by the angle α subtended at the observer's eye (unaided by the microscope) by the object at this same distance D. Referring to Fig. 3.8, where the angles α and α' are small, and assuming the eye is close to the eyepiece, the angular magnification is given by:

$$m_A = \frac{\alpha'}{\alpha} = \frac{h_3/D}{h_1/D} = \frac{h_3}{h_1}$$

For the objective lens system,

$$\frac{1}{u} + \frac{1}{v} = \frac{1}{f_o}$$

where f_o is the focal length of the objective, and

$$\frac{h_2}{h_1} = \frac{v}{u} = \frac{v}{f_o} - 1$$

For the eyepiece lens system, where the emergent vergence is negative,

$$\frac{1}{u'} - \frac{1}{D} = \frac{1}{f_E}$$

and

$$\frac{h_3}{h_2} = \frac{D}{u'} = 1 + \frac{D}{f_E}$$

Therefore,

$$m_A = \frac{h_3}{h_1} = \left(\frac{h_3}{h_2}\right)\left(\frac{h_2}{h_1}\right) = \left(1 + \frac{D}{f_E}\right)\left(\frac{v}{f_o} - 1\right) \qquad (3.5)$$

For 'normal' adjustment, the final image is at infinity and the intermediate image is at the first focal point of the eyepiece. Hence, $u' = f_E$ and

$$m_A = \frac{\alpha'}{\alpha} = \frac{h_2/f_E}{h_1/D} = \frac{h_2 D}{h_1 f_E} = \left(\frac{v}{f_o} - 1\right)\frac{D}{f_E} \qquad (3.6)$$

which is less than m_A in the arrangement leading to Equation 3.5.

As the focal length f_o of the objective is often much less than v (the separation $P'_o F_E$ when the adjustment is normal), Equation 3.6 can be written as:

$$m_A = \frac{vD}{f_o f_E} \qquad (3.7)$$

Thus, large angular magnification (and hence large linear magnification) for a given value of D is obtained by a large value of v—which requires a long microscope tube—but particularly by the choice of a small value of f_o and, to a lesser extent in practice, of f_E.

The passage of a typical cone of rays through a compound microscope when in normal adjustment is shown in Fig. 3.9.

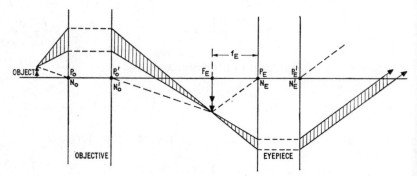

Fig. 3.9. Passage of a typical cone of rays through a compound microscope in normal adjustment

3.6 Microscope Objectives

For a microscope, the object is small and is at a relatively small distance from the objective; thus, the angular field of view is small but large-angle cones of rays will pass through the instrument from object points. This is particularly so for high-power objectives, when the object plane may only be a few millimetres from the nearest surface of the objective. The objective must therefore be carefully corrected for spherical aberration and coma, whereas the field-dependent aberrations are of less importance. Chromatic aberration must also be corrected as any appreciable amount present would give rise to an offensive image.

A single plano-convex doublet of the Lister type (Section 2.4) with a focal length of several centimetres is suitable for a low-power objective (for example, for a travelling microscope, for which objective focal lengths of 1 in and 3 in are common). The object plane will then be one real Lister plane, and the interme-diate image plane the other Lister plane, as in Fig. 3.10 where L and L' are the two real Lister points.

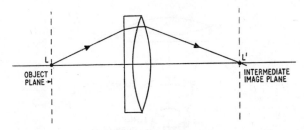

Fig. 3.10. Lister doublet as a low-power microscope objective

A medium-power objective may comprise two or more doub-lets and is a simple extension of the above principle. The Lister point L_1 of the first doublet (doublet 1 in Fig. 3.11) will be in the object plane. The image Lister point L_1' of doublet 1 will be virtual and will form the real object Lister point of doublet 2; the other real Lister point of doublet 2 is at L_2', in the plane of

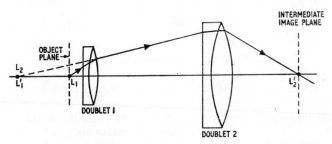

Fig. 3.11. Two Lister doublets forming a medium-power microscope objective

which the intermediate image lies. This design will be suitable for focal lengths of 1 cm or 2 cm and numerical apertures (Sec-tion 12.4) of about 0·15.

With high-power objectives, a modified form of the Lister arrangement is used in which the first element is a hemispherical lens with the necessary focal length of a few millimetres. This component will, of course, introduce considerable chromatic and spherical aberration which must be compensated for in the remaining components. A highly corrected objective of this nature is called an *apochromatic objective;* the numerical aperture of such an objective will be in the region of 0·8.

To achieve a significant increase in numerical aperture, an *oil-immersion objective* must be used; a typical design is shown in Fig. 3.12. A_1 is the object (point of detail in a microscope slide)

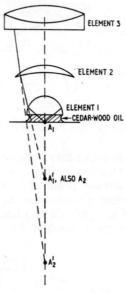

Fig. 3.12. An oil-immersion objective

immersed in cedar-wood oil, which is placed between the slide and the lower surface of the first component of the objective. The diagram is greatly exaggerated: only a film of oil will be concerned. The refractive index of cedar-wood oil is 1·516 (for the wavelength of the sodium D line), which is the same as that of the hemispherical component (element 1) of the objective. Thus, by adjustment of the objective position, the object may be brought to an aplanatic point because it is effectively embedded in the first component (Section 2.9). The image of A_1 formed by the element 1 will be at A_1', and spherical aberration will be minimised for this element because A_1' is the second aplanatic point. A_1' is also chosen to be the centre of curvature of the first surface of the meniscus lens (element 2 of the objective), so that the rays from element 1 will pass normally through this first surface and will incur no further spherical aberration until the second surface of element 2 is reached. This second surface of element 2 is given a suitable radius so that the point A_1' is also A_2, the first aplanatic point of a sphere having this radius and the refractive index of element 2. The image, after refraction at element 2, will be at A_2', the second aplanatic point of the sphere associated with element 2. This point A_2' may now form the first of a series of Lister points for the succeeding elements. By this means, the numerical aperture of the objective may significantly exceed unity.

Reflecting objectives for microscopes are described under interference microscopy, and the applications of infra-red and ultraviolet light, in Volume 2.

3.7 Eyepieces (Oculars)

The main functions of an eyepiece are, briefly, to magnify the image produced by the objective and to make full use of the limit of resolution of the objective.

The *limit of resolution* of an optical system is the angular separation θ between two points in an object which can be discerned as separated in the image of the object. This limit of resolution, or *resolving power*, is related to the wavelength λ of the light transmitted from the object through the system to the image and depends on a criterion (for example, the *Rayleigh criterion*) as to when the detail can be regarded as resolved. This question of resolution is considered more fully in Chapter 12. For a circular aperture of diameter D—and, as can be shown, for an objective lens of this diameter — the limit of resolution is given by:

$$\theta = \frac{1.22\lambda}{D} \tag{3.8}$$

The limit of resolution θ_E of the human eye is correspondingly related to its entrance pupil diameter; for light of wavelength 5,550 Å, at which visual sensitivity is a maximum (Section 6.7), θ_E is approximately 3×10^{-4} rad.

If, for example, an astronomical telescope has an objective of diameter 20 cm, its limit of resolution for $\lambda = 5,500$ Å is:

$$\theta = \frac{1.22 \times 5.5 \times 10^{-5}}{20} = 3.4 \times 10^{-6} \text{ rad}$$

The optimum magnifying power of this telescope is consequently:

$$\frac{\theta_E}{\theta} = \frac{3 \times 10^{-4}}{3.4 \times 10^{-6}} = 88$$

With this objective, an attempt to achieve an angular magnification m_A larger than 88 by choice of f_o, the objective focal length, and f_E, the eyepiece focal length, so that f_o/f_E exceeds 88 will be frustrated by the fact that no greater detail is resolvable.

It is difficult to construct a satisfactorily corrected eyepiece of focal length less than a few centimetres. If the focal length of the eyepiece is, say, 2·5 cm, the optimum focal length of the objective in this example will be $88 \times 2·5 = 220$ cm. However, the optimum magnification will clearly increase with larger objective diameter (θ decreases in Equation 3.8 and hence θ_E/θ increases), and there will be the added advantage of increased light-gathering power for the objective.

The choice of the magnifying power of the eyepiece alone must not be such that the diameter of its exit pupil or *Ramsden circle* is greater than the mean diameter of the eye pupil, otherwise the iris itself will form the aperture stop of the system and the resolution limit will be reduced. In the example given above, the diameter of the Ramsden circle will be 1/88 times that of the objective, i.e. $20/88 = 0·23$ cm, which is well within the mean diameter of the eye pupil, so no restriction arises.

Apart from the residual aberrations due to the objective, the eyepiece itself must be carefully corrected for the field-dependent aberrations—i.e. for astigmatism, curvature of the field and distortion (see Table 2.1)—because the short focal length results in a large angle of view.

The simplest eyepiece usually comprises a single converging lens or diverging lens or a doublet. This design allows only a limited degree of correction and will not be considered further here.

The commonest eyepiece lens systems employ two separated elements: the element nearest the objective in the optical instrument (telescope or microscope) is called the *field lens;* the other element nearest the eye is called the *eye lens.* The function of the field lens is to increase the angular field of view compared with what this field would be if the eye lens were used alone. This is illustrated in Fig. 3.13, where Ω is the semi-angular field of view without the field lens, and Ω' is that obtained with the field lens. The function of the eye lens is to magnify the image produced by the objective. Furthermore, the eye lens is combined with the field lens in such a way as to reduce aberration. The chief two-element systems used in this connection are the Huygens' eyepiece and the Ramsden eyepiece.

The *Huygens' eyepiece* consists of two thin plano-convex lenses, of which the field lens L_F has a focal length three times

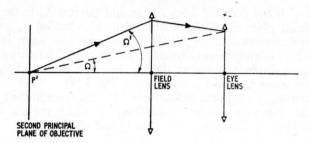

Fig. 3.13. *The semi-angular field of view in relation to the use of the field lens*

that of the eye lens L_E (Fig. 3.14); the separation between the two lenses is twice the focal length of the eye lens. The plane surfaces of the two lenses are towards the eye. The focal length of this combination is found from Equation 1.20 for $_1n_2 = 1$ (since

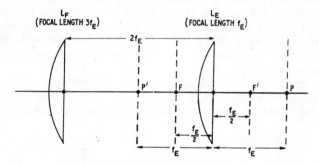

Fig. 3.14. *The Huygens' eyepiece*

the lenses are in air), $_1d_2 = 2f_E$ (where f_E is the focal length of the eye lens), $F_1 = 1/3f_E$ and $F_2 = 1/f_E$. Therefore,

$$\frac{1}{f} = \frac{1}{3f_E} + \frac{1}{f_E} - \frac{2f_E}{3f_E^2} = \frac{2}{3f_E}$$

i.e. the focal length f of the combination is $3/2$ times that of the eye lens.

Again, in Equation 1.21, the axial distance f' between the second principal point P' and the second focal point is $3f_E/2$; the distance f_1' from the first element to its own second focal point is $3f_E$; and $_1d_2 = 2f_E$. Therefore,

$$v' = \frac{3f_E}{2}\left(\frac{3f_E - 2f_E}{3f_E}\right) = \frac{f_E}{2}$$

where v' is the distance from the second element L_E to the second focal point F' of the combination. F' is behind L_2 on the side towards the eye. As P', the second principal point of the combination, must be at a distance $3f_E/2$ before F', it follows that P' is at a distance f_E before the eye lens (i.e. midway between the two lenses L_F and L_E, as shown in Fig. 3.14).

Likewise, as suggested in Section 1.11, a ray-trace in the opposite direction gives the first focal length f, which must also be $3f_E/2$. Then in Equation 1.21, $f'_1 = f_E$ and $v' = v$; so

$$\frac{3f_E}{2} = \frac{vf_E}{f_E - 2f_E}$$

Therefore,

$$v = -\frac{3f_E}{2}$$

which is the separation between the field lens L_F and the first focal point F. Hence, F is at a distance $3f_E/2$ from the field lens towards the eye lens, i.e. in the opposite direction to the direction of the ray-trace. As $PF = 3f_E/2$, the first principal point must be separated from the eye lens by a distance f_E towards the eye, as shown in Fig. 3.14.

The deviation of an incident light ray produced at the field lens of a Huygens' eyepiece is equal to that produced at the eye lens, so the deviation in the system is equally distributed between the

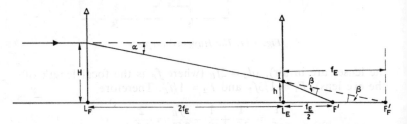

Fig. 3.15. In the Huygens' eyepiece deviation is equally shared between the two elements so spherical aberration is a minimum

two lenses; this gives minimum spherical aberration (see Section 2.4). To show that this is so, consider an incident ray parallel to the optical axis which meets the field lens L_F at a separation H from the axis (Fig. 3.15). After refraction by the lens, this ray would (if the eye lens L_E were absent) cut the optical axis at F'_F, the second focal point of L_F, which is at an axial distance $3f_E$ from L_F. The angle of deviation α of this ray at L_F

is, therefore, equal to $H/3f_E$, provided that $H \ll 3f_E$. In fact, this ray is deviated again, at the eye lens L_E, through an angle β and intersects the optical axis at F', the second focal point of the combination, which is at a distance $f_E/2$ beyond L_E. If the separation of the point of incidence I on L_E from the axis is h, it follows that $\beta = h/f_E$ because $\angle IF'_F F' = \angle F'IF'_F = \beta$ if h is small compared with f_E. From similar triangles,

$$\beta = \frac{h}{f_E} = \frac{H}{3f_E} = \alpha$$

The Huygens' eyepiece is also corrected for lateral (but not longitudinal) chromatic aberration because the separation between the field lens and the eye lens equals the mean of the individual focal lengths, in accordance with Equation 2.17. However, to obtain really satisfactory freedom from chromatism, each element must be corrected separately, i.e. both the field and eye lens need to be achromatic doublets.

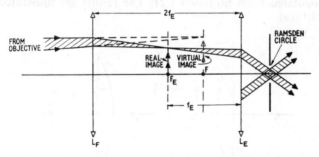

Fig. 3.16. *Passage of a cone of rays from the objective through a Huygens' eyepiece in a low-power microscope in normal adjustment*

Fig. 3.16 illustrates, for a low-power microscope in normal adjustment, the passage of a typical cone of rays from the objective through a Huygens' eyepiece. For the light to emerge from the eye lens in parallel rays, the image formed by the objective, i.e. the intermediate image in the instrument, should be in the first focal plane of the eyepiece. For a Huygens' eyepiece, the first focal plane through F is between the field and eye lenses (Fig. 3.14). As shown in Fig. 3.16, therefore, the cone of rays from the objective is refracted by the field lens L_F so that it converges to a point in an intermediate image before reaching the focal plane through F. If it is to be rendered parallel on emergence from the eye lens L_E, this intermediate image must now be in the first focal plane of the eye lens only, through F_E.

An eyepiece of which the first focal plane is between the lenses is called a *negative eyepiece*. It has the disadvantage that any cross-wires or graticule scale used has to be placed between the two elements. If these cross-wires are placed in the plane of the intermediate image, their image will be formed by the eye lens L_E only and will suffer aberration unless L_E itself is a highly corrected element. In practice, Huygens' eyepieces are designed in which the lens separation departs somewhat from $2f_E$.

The *Ramsden eyepiece* comprises two plano-convex lenses of equal focal length f, usually separated by a distance $2f/3$, and with the convex surfaces facing one another. The focal length f_R of this combination of lenses is given by Equation 1.20:

$$\frac{1}{f_R} = \frac{2}{f} - \frac{2f/3}{f^2} = \frac{4}{3f}$$

Therefore, $f_R = 3f/4$. The positions of the principal points can be calculated from Equation 1.21. The results are illustrated in Fig. 3.17(a).

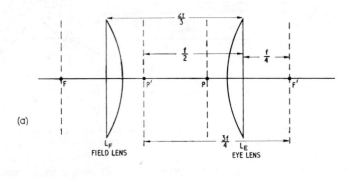

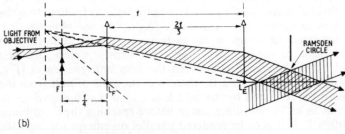

Fig. 3.17. The Ramsden eyepiece: (a) the cardinal points; (b) passage of a conc of rays through the eyepiece

The focal planes are in the region outside the lens elements. Consequently, the intermediate image formed by the objective is now in the first focal plane of the eyepiece and in front of this eyepiece when the instrument is in normal adjustment [Fig. 3.17(b)]. The Ramsden eyepiece is, therefore, a *positive eyepiece*, and any cross-wires (or graticule scale) can be placed in the plane of the intermediate image and observations made through the eyepiece without difficulty of focusing the image on the cross-wires. This is a great advantage when object dimensions are to be measured against a microscope or telescope scale.

The Ramsden eyepiece does not satisfy the condition of Equation 2.17 that, for freedom from lateral chromatic aberration, the separation between the lenses should be the mean of their focal lengths. For this condition to hold, the separation between the lenses would need to be f. This would be unsatisfactory as the field lens surface would then be visible in the final image. The usual choice of a separation of $2f/3$ is a compromise. This does not give as good freedom from spherical aberration as the Huygens' eyepiece, but the advantages of a positive eyepiece are considerable. The Huygens' and Ramsden eyepieces both use plano-convex lenses which are designed to compensate as far as possible for the highly field-dependent aberrations.

The main developments from the basic Ramsden and Huygens' eyepieces have been aimed at correcting the chromatic defects of the Ramsden eyepiece and increasing the field of view. These improvements usually result in two or more elements for the eye lens. Many very complex eyepiece designs not based on the Ramsden and Huygens' types also exist but their details are beyond the present scope.

3.8 Depth of Field and Depth of Focus

Depth of field is concerned with the effect on the observed image formed by an optical system of a given object when the object is moved; depth of focus is concerned with movement of the image plane.

The eye has a certain degree of tolerance such that it will accept as a point a circular patch of light of small enough size which is sometimes called a *blur circle*. If the radius of a blur circle is stipulated, a point object may be moved along the axis of an optical system between two extreme positions where the image in a fixed observing plane is within the radius of the blur circle and so can still be considered in sharp focus. For example, in Fig. 3.18, the point object may be moved from A to B through a distance D along the optical axis whilst the conjugate image moves from A' to B' through a distance L, the extreme positions giving blur circle images of critical radius ϱ (exaggerated in the diagram). The distance $A'B' = L$ (the depth of field) will be seen from Fig. 3.18 to increase as the radius R of the exit pupil decreases. The relation between R and L is not a simple one.

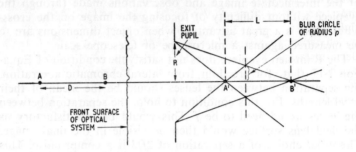

Fig. 3.18. Depth of field and the exit pupil radius

If the object position is fixed at A and the observing plane moved, the situation is as illustrated in Fig. 3.19. Here, the depth of focus d is the distance along the optical axis that the image plane can be moved between the critical blur circles at C and E; again this distance increases with decrease of the exit pupil radius.

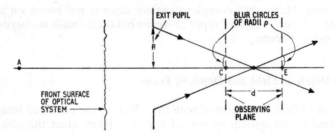

Fig. 3.19. Depth of focus and the exit pupil radius

3.9 The Camera

The photographic camera consists essentially of a converging objective lens of focal length f and exit pupil aperture diameter d (and hence f-number $N = f/d$) which produces a real image on a flat film or plate of an object. If this object is at infinity (if, for example, photography of a distant landscape is undertaken), the film or plate is in the second focal plane of the objective lens. In order to obtain a focused image at the film of nearer objects, the objective has to be moved along the optical axis further away from the second focal plane. The objective–film separation is, therefore, usually variable at will on focusing, a facility made available by the arrangement of extensible light-tight bellows between the objective mount and the film or plate holder or, as in most modern miniature camera practice, by means of a light-tight metallic cylinder in which the objective is mounted and which can be moved on a helical thread relative to a fixed cylinder attached to the camera body.

The angular width of the field, and hence the area of a given object field recordable, will depend on the focal length of the objective and the dimensions of the area of film or plate on which the image is formed. The roll films most frequently used in modern cameras are the 35 mm films which give a picture area of 35 mm×24 mm, and the sizes known as 620 and 120 where the picture area is usually $2\frac{1}{4}$ in ×$2\frac{1}{4}$ in. Larger cameras employing plates or film packs may be full-plate size ($8\frac{1}{2}$ in×$6\frac{1}{2}$ in), half-plate ($6\frac{1}{2}$ in×$4\frac{3}{4}$ in), quarter-plate ($4\frac{1}{4}$ in×$3\frac{1}{4}$ in) or 4 in×5 in.

The normal-focus lens for a camera has a focal length f about equal to the diagonal of the picture area. The normal objective must therefore be capable of an angle of view such as to provide a flat field over a circle of diameter f. This corresponds to an angle of view θ (Fig. 3.20) given by:

$$\theta = 2 \tan^{-1} 0\cdot5 = 53°$$

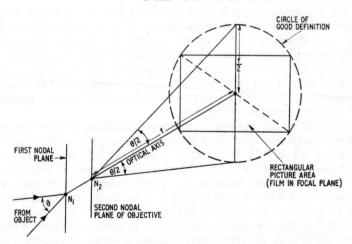

Fig. 3.20. Angle of view of a camera

This corresponds approximately to the angle which the eye can cover without the head being moved.

The film area exposed when a scene is recorded is usually rectangular. It is clear from Fig. 3.20 that, with the long side of the rectangle horizontal, the angle of view in the horizontal plane is less than θ and in the vertical plane it is less still. A miniature 35 mm camera will thus have a normal objective lens of focal length $\sqrt{(35^2 + 24^2)} = 42\cdot5$ mm for an angle of view of 53°. Usually, a 50 mm focal length lens is normal, giving an angle of view of $2 \tan^{-1} (21\cdot25/50) = 45°$. A 4 in×5 in plate camera will have a normal objective of focal length $\sqrt{(4^2 + 5^2)}$ in, i.e. between 6 in and 7 in.

Bellows-type cameras often have the facility for double or triple extension beyond the normal objective focal length so that lenses of longer focal length can be employed instead of the normal one; this clearly narrows the field of view. Alternatively, a lens of shorter focal length than normal can be used with the bellows retracted, providing a wide-angle facility.

Many modern miniature 35 mm cameras can be fitted with interchangeable lenses of different focal lengths. For example, if such a camera has a normal objective of focal length 50 mm and an angle of view of 45°, this lens can be interchanged with a telephoto lens of focal length 135 mm (angle of view 18°) or other sizes, such as 250 mm (angle of view 10°) or even 1,000 mm (angle of view 2·5°). Alternatively, replacement by a wide-angle lens of focal length 35 mm provides an angle of view of 62°, whilst one of 25 mm gives an angle of 82°.

On a simple basis, it is clear that the light-passing power of a camera objective will be proportional to its aperture area (and hence to d^2, where d is the aperture diameter); it is also inversely proportional to the square of the lens–film separation, so proportional to $1/f^2$ when the camera is focused at infinity, f being the focal length of the objective. The light-passing power is hence proportional to d^2/f^2, i.e. inversely proportional to the square of N, the f-number of the lens.

To give a high light-passing power, therefore, the lens must have a small f-number. In miniature cameras, $f/2$ is often available; in $2\frac{1}{4}$ in×$2\frac{1}{4}$ in cameras, $f/4\cdot5$ to $f/2\cdot8$; and in larger cameras, $f/4\cdot5$ to $f/6\cdot3$. An $f/2$ lens of focal length 50 mm for a 35 mm camera will have an exit pupil diameter of 25 mm. A camera for a 4 in×5 in film with a 7 in focal length objective would need to have a lens of exit pupil diameter $3\frac{1}{2}$ in to attain $f/2$. One of the reasons for the advance in miniature cameras has been the design of large-aperture lenses of small f-numbers and short focal lengths, demanding relatively small sizes of high-quality optical glass blanks for fabrication; such lenses are available with long focal lengths but are extremely expensive.

It is clear from Section 3.8 that the depth of field and depth of focus are both increased as the objective aperture diameter is reduced. Camera objectives are provided with an iris diaphragm of variable aperture so that this reduction can be readily achieved. This diaphragm is usually between the components of the compound objective lens. Frequently, its operating mechanism is provided with click stops which, for a full aperture of $f/2$, will give successive reductions to, for example, $f/2\cdot8$, $f/4$, $f/5\cdot6$, $f/8$, $f/11$ and $f/16$. It is easily seen that each reduction from one to the next smaller stop in this series approximately halves the light-passing power of the lens. For example, a change from $f/5\cdot6$ to $f/8$ corresponds to a reduction in light-passing power of $5\cdot6^2/8^2 = \frac{1}{2}$ approx. Similarly, for the same object and degree of illumination, the exposure time must be doubled.

To vary the exposure time, a *shutter* is used. In a *leaf-type shutter* (of which the Compur shutter is well known), this takes the form of interleaving thin metallic blades usually between the objective lens components immediately in front of the iris diaphragm. In a *focal-plane shutter*, a double blind of fabric or **metal** is arranged to pass the front surface of the film or plate; as the blind passes the surface, a slit between the two blind components, the width of which can be varied, is drawn across. Focal plane shutters are almost invariably used for miniature cameras with interchangeable lenses to avoid providing each separate lens with its own leaf-type shutter. The shutter opening times are arranged by a clockwork mechanism and are generally 1s, $\frac{1}{2}$ s, $\frac{1}{4}$ s, $\frac{1}{8}$ s, $\frac{1}{15}$ s, $\frac{1}{30}$ s, $\frac{1}{60}$ s, $\frac{1}{125}$ s, $\frac{1}{250}$ s and $\frac{1}{500}$ s, with provision also for longer times measured by a stopwatch.

The many types of cameras and their operation are beyond the present scope, as also are the topics of speed, sensitivity, characteristics of films and plates and also processing. The brief account given concerns only the essential chief features common to all cameras as optical instruments. For further information, reference should be made to the Manuals of Photography published by the larger photographic firms such as Ilford Ltd and Kodak Ltd in Great Britain.

3.10 Camera Objectives

The simplest type of camera objective used, for example in the 'box camera', is the 'landscape' type in the form of a single biconvex lens with a single aperture or limited range of apertures.

The spherical aberration is minimised by means of a glass of refractive index 1·5, and a radius of curvature of the back surface six times that of the front surface (Section 2.4). Such a lens has considerable field curvature, so only the image at the centre of the flat plate or film in the camera is sharply focused. The Wollaston type of meniscus lens, with the concave surface of longer radius towards the object and a stop at the centre of curvature of the convex back face, was introduced in 1812 and gave less field curvature and coma. Even so, the maximum aperture possible is about $f/16$. To reduce chromatic aberration, achromatic doublets, especially of meniscus form, were introduced and later, two separated achromatic doublets.

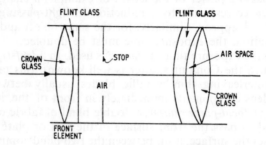

Fig. 3.21. The Petzval portrait lens

A considerable advance in design was achieved in 1840 by Petzval, who designed a portrait lens (Fig. 3.21) of $f/4$ which, however, exhibited astigmatism and field curvature. It comprised two separated achromatic combinations: the front one (towards the object) being a cemented doublet, and the back doublet containing a meniscus-shaped air space. This lens was the first one computed mathematically in accordance with Petzval's condition that the sum of the refractive indices of the individual lenses multiplied by the reciprocals of their powers should be zero, i.e. $\Sigma(n/F) = 0$. It was later improved by Dallmeyer.

Several advances took place subsequently but the most notable one was in 1886. Then, the introduction of a greater variety of optical glasses by Schott and Genossen in Jena, Germany, enabled lenses with small astigmatism to be designed by combining lens components of high refractive index and low dispersion with ones of low refractive index and high dispersion. Such lens designs, known as 'anastigmats' were widely used in Germany and Britain.

The Cooke lens, designed by H. Dennis Taylor and introduced in 1893, was a considerable simplification and advance on the previous anastigmats and employed initially only three separated elements (Fig. 3.22). The two outer biconvex elements were of high-refractivity low-dispersion glass, shaped for minimum spherical aberration; between them was a symmetrical biconcave lens of light flint glass which flattened the field considerably. An iris diaphragm stop between the biconcave and back biconvex lens was introduced and provided a maximum aperture of $f/4\cdot5$, later increased to $f/3\cdot5$.

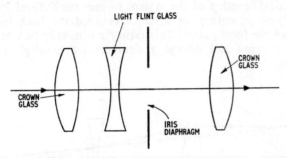

LIGHT FLINT GLASS

CROWN GLASS

CROWN GLASS

IRIS DIAPHRAGM

Fig. 3.22. The Cooke triplet objective lens

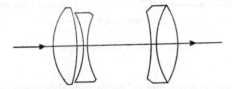

Fig. 3.23. The Tessar lens

In 1902, Rudolph designed the Zeiss Tessar lens (Fig. 3.23) which has similarities with the Cooke lens although a cemented meniscus doublet was used for the back component. The Tessar and other Zeiss lenses were really the fore-runners of the remarkable range of modern photographic objectives available, benefited by the recent introduction of rare earth glasses with particularly high refractive index and low dispersion. The design of the modern objectives, with superb resolution and freedom from aberration over an almost flat field, is a complex science now involving the digital computer as an aid to the calculation of light paths through the lens. Six or more components are not uncommon, with the element surfaces 'bloomed' to improve light transmission and reduce back-reflection (Section 9.4).

The *telephoto lens* comprises a converging front component followed by a widely separated diverging back component. As can be seen from Fig. 3.24, the effect of the diverging element is to increase the equivalent focal length of the system, whilst maintaining a comparatively small separation between the front element and the second focal plane of the lens. The principal planes are well in front of the lens, so the camera extension is much less than would be needed if a long-focus ordinary objective was used. The principle is essentially the same as in the Galilean telescope (Section 3.3). The 'telephoto effect' is the ratio of the focal length f of the system to the 'back' focal length v (the bellows extension, or distance between the back lens element and the focal plane). This ratio f/v is usually two or three. The converging and diverging elements are usually corrected doublets.

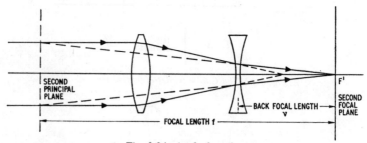

Fig. 3.24. A telephoto lens

It is possible to have one or more movable components in an objective so as to provide a variable focal length and consequently a variable angular field. The combination of the characteristics of wide-angle normal-focus components and telephoto components in a single objective, known as a *zoom lens*, is popular in cinematography and in miniature 35 mm camera still photography.

Exercise 3

1. Define the cardinal points of a thick lens system.
   The cornea of a human eye has a radius of curvature of 0·78 cm and the distance from the cornea to the retina is 2·5 cm. The material of the eye (apart from the crystalline lens) has a refractive index of 4/3. If the crystalline lens may be treated as a thin lens 0·5 cm behind the cornea, what focal length (measured *in situ*) must it have in order that the eye may focus on an object at infinity? Find the principal focal lengths of the system. (L. P.)

2. A telescope consists of two thin converging lenses whose focal lengths have numerical values F and f ($F > f$) at a distance $F+f$ apart. Find the relation between the object position and the final image position, for any position of the object. Show that if the object is moved a distance s further away, the image moves a distance sf^2/F^2. Show also that the lateral magnification of the image is f/F, and comment on the fact that the apparent magnification is greater than unity. (L. Anc.)

3. Give an account of either (a) the chromatic aberration of a coaxial spherical lens system, or (b) two monochromatic aberrations of importance in a telescope. Explain how the aberrations you describe may be corrected in a telescope. (L. G.)

4. Describe the structure and explain the mode of action of a modern microscope. (L. Anc.)

5. Write an essay on the optical design and performance of either (a) the camera, or (b) the optical microscope. (L. P.)

6. Discuss two of the defects in the image formed by a single lens and the means used to correct them.
What are aplanatic points with respect to a single spherical refracting surface? Explain the use made of such points in the design of the oil immersion objective. (L. Anc.)

7. Find the positions of the cardinal points in an eyepiece consisting of two thin converging lenses each of focal length 4·0 cm and separated by a distance of 3·0 cm. The eyepiece is used in an astronomical telescope in normal adjustment. The diameter of the objective is 5 cm and its focal length is 60 cm. Find the position and diameter of the exit pupil. (L. G.)

8. An eyepiece consists of two coaxial thin plano-convex lenses of focal length 6 cm and 2 cm respectively, separated by 4 cm and with the shorter focus lens and the plane surface nearer the eye. Determine the positions of the cardinal points of the eyepiece. Assuming that the material of the two lenses is the same, show that the focal length is the same for all colours and mention one other feature of this eyepiece. (L. G.)

9. A telescope eyepiece consists of two thin converging lenses, a field lens of focal length 6 cm and an eye lens of focal length 2 cm, separated by a distance of 4 cm. Calculate the positions of the cardinal points of the system and use them to construct a diagram, drawn to scale, to illustrate the action of the eyepiece. State, giving reasons, where the cross wires should be situated, and discuss the chromatic aberration of the eyepiece if the two lenses are of the same glass. (L. P.)

10. State the properties associated with the principal (or unit) planes of a coaxial lens system and describe how the positions of the planes can be found experimentally. What special features are required in a telephoto lens system?

A telephoto lens consists of a converging lens and a diverging lens of focal lengths 5·0 and 2·0 cm respectively placed coaxially 3·5 cm apart. At what distance from the diverging lens will the image of a distant object be formed and what will be the size of this image if the object subtends an angle of 5°? (L. G.)

11. Explain what is meant by (a) the numerical aperture of a microscope objective, (b) the f-number of a camera lens. Discuss the importance of these quantities in the performance of these instruments. (G. I. P.)

12. Discuss the design parameters for a compound microscope and show how the magnification may be (a) calculated and (b) measured. Mention briefly some of the recent advances in the field of microscopy. (G. I. P.)

13. A Ramsden eyepiece consists of two thin converging lenses each of focal length 3 cm, separated by a distance of 2 cm. Determine the positions of the cardinal points of the system. The cardinal points should be defined and their position, for the eyepiece, shown on a diagram.
Comment on the aberrations of the instrument. (G. I. P.)

14. Explain why most visual optical instruments are fitted with eyepieces consisting of two or more lenses rather than with a single lens.
An eyepiece consists of two thin converging lenses each 2 cm in diameter. The field lens has a focal length of 4 cm, and the eye lens 2 cm, and their separation is 3 cm. Find the cardinal points. The eyepiece is used in a telescope with an objective 3 cm in diameter and 15 cm focal length Find the position and size of the Ramsden disc or eye-ring and the angular magnification of the system in normal adjustment. Illustrate your answer with a full-scale labelled diagram. (G. I. P.)

CHAPTER 4

Refractometry

This chapter is concerned with the measurement of the refractive indices of solids and of liquids which are transparent to visible light. Procedures involving interference methods (for example, for thin films and for gases) are deferred until Chapters 10 and 11. The measurement of the refractive indices of solids which are opaque to light can sometimes be undertaken by methods involving the use of polarised light: this is considered in Chapter 3, Volume 2.

4.1 Determination of Refractive Indices of Solids and Liquids by Prism Methods

The measurement of the angle of deviation which a beam of light undergoes on passing through a prism can be used to calculate the refractive index of the prism material. Elementary procedures which make use of rectangular prisms or blocks of the material are not described here. More accurate procedures utilise a prism of triangular cross-section. For incident monochromatic light, the angle of deviation on passage through the prism varies with the angle of incidence on the first face. If the refractive index of the material of the prism is to be measured, the exact position of the prism relative to the incident beam must therefore be known.

A simpler and preferable procedure is, however, to determine the angle of minimum deviation D of a beam of light on passage through the prism and to calculate n, the refractive index of the prism material relative to air taken to be unity, from the well-known equation

$$n = \frac{\sin\left[(A+D)/2\right]}{\sin A/2}$$

where A is the refracting angle of the prism.

A narrow collimated beam of monochromatic light (provided, for example, by a sodium lamp as the source against the slit of a collimator of a spectrometer) is passed through a triangular prism; I_1 is the angle of incidence at the front face, R_1 is the angle of refraction, I_2 is the angle of emergence at the second face and R_2 is the angle of refraction (Fig. 4.1). The angle of deviation D has a minimum value for a certain value of I_1 or I_2 when dD/dI_1 or dD/dI_2 is zero. Now,

$$\sin I_1 = n \sin R_1$$

and

$$\sin I_2 = n \sin R_2$$

Therefore,

$$\cos I_1 \, dI_1 = n \cos R_1 \, dR_1$$

and

$$\cos I_2 \, dI_2 = n \cos R_2 \, dR_2$$

Therefore,

$$\frac{dI_1}{dI_2} = \frac{\cos R_1 \, dR_1 \cos I_2}{\cos R_2 \, dR_2 \cos I_1}$$

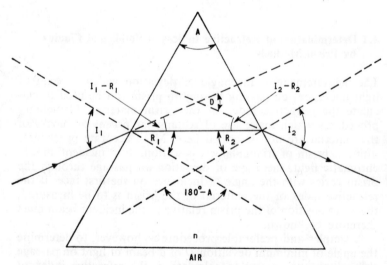

Fig. 4.1. *Passage of a narrow collimated beam of monochromatic light through a triangular prism*

But

$$A = R_1 + R_2$$

Hence,

$$R_2 = A - R_1$$

and

$$dR_2 = -dR_1$$

because the refracting angle A is constant. Therefore,

$$\frac{dI_1}{dI_2} = -\frac{\cos R_1 \cos I_2}{\cos R_2 \cos I_1}$$

The angle of deviation D is given by the equation:

$$D = I_1 + I_2 - A$$

Therefore,

$$\frac{dD}{dI_2} = \frac{dI_1}{dI_2} + 1$$

or, for a minimum value,

$$\frac{dI_1}{dI_2} = -1$$

Thus, for a minimum value,

$$\frac{\cos R_1 \cos I_2}{\cos R_2 \cos I_1} = 1$$

This relation is satisfied when $I_1 = I_2$ and $R_1 = R_2$, i.e. when the light passes symmetrically through the prism. It could not be satisfied for $I_1 = R_1$ because then there would be no refraction.

Let $I_1 = I_2 = I$ and $R_1 = R_2 = R$; it can then be seen from Fig. 4.1 that, at minimum deviation,

$$D = (I_1 - R_1) + (I_2 - R_2) = (I_1 + I_2) - (R_1 + R_2) = 2(I - R)$$

and

$$A = R_1 + R_2 = 2R$$

Therefore,

$$D + A = 2I$$

and

$$n = \frac{\sin I}{\sin R} = \frac{\sin \left[(D+A)/2 \right]}{\sin A/2} \qquad (4.1)$$

A good-quality triangular prism of the glass or transparent material and with optically polished faces meeting at the angle A (usually 60°) is set on a spectrometer table. The angle A is measured and then the angle of minimum deviation D, usually with light of wavelength 5,893 Å from a sodium lamp. The refractive index n is calculated from Equation 4.1.

This method is also useful for determining the refractive index of a transparent liquid available in sufficient quantity to fill a hollow triangular prism made from thin plane-parallel sheets of glass cemented together. When empty, the hollow prism produces no deviation of the light. If, however, the prism sides are not optically flat, the observed image of a spectrometer slit may be slightly diffuse.

The difference in refractive index between two liquids may be determined by measuring the angle of deviation of a beam of collimated monochromatic light which traverses a cell of square cross-section provided with a diagonal partition [Fig. 4.2(a)]. The walls of the cell and the partition are made of optically flat glass. One half of the cell contains a standard solution (often distilled water), and the other half contains another transparent liquid or solution of which the refractive index is required. A beam of monochromatic light from the collimator of a spectrometer enters normally one face of the cell mounted on the spectrometer table; the angle of deviation is measured with the telescope. If the telescope can resolve objects 2 cm apart at a distance of 1 km, corresponding to an angle of 2×10^{-5} rad or $(1 \cdot 15 \times 10^{-3})°$, a difference of refractive index of about 1×10^{-5} may be detected.

To evaluate $\varDelta n$, the difference between the refractive indices n_1 and n_2 in Fig. 4.2(a), use can be made of the equation:

$$\frac{n_2}{n_1} = \frac{\sin 45°}{\sin x}$$

Therefore,

$$\sin x = \frac{n_1}{n_2 \sqrt{2}}$$

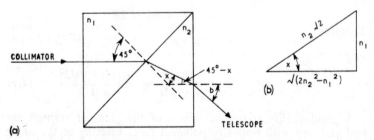

Fig. 4.2. Method of determining the difference in refractive index between two liquids

which leads to Fig. 4.2(b). Also, from Fig. 4.2(a),

$$n_2 = \frac{\sin b}{\sin (45° - x)} = \frac{\sin b}{\sin 45° \cos x - \cos 45° \sin x}$$

Hence, reference to Fig. 4.2(b) shows that:

$$n_2 = \frac{\sin b}{(1/\sqrt{2})[\sqrt{(2n_2^2 - n_1^2)}/n_2\sqrt{2}] - (1/\sqrt{2})(n_1/n_2\sqrt{2})}$$

$$= \frac{2n_2 \sin b}{\sqrt{(2n_2^2 - n_1^2)} - n_1}$$

Therefore,

$$\sqrt{(2n_2^2 - n_1^2)} = n_1 + 2 \sin b$$

Hence,

$$2n_2^2 - n_1^2 = n_1^2 + 4n_1 \sin b + 4 \sin^2 b$$

or

$$n_2^2 - n_1^2 = 2n_1 \sin b + 2 \sin^2 b \tag{4.2}$$

If $n_2 = n_1 + \Delta n$ and Δn is small,

$$n_2^2 = n_1^2 + 2n_1 \Delta n + (\Delta n)^2 \tag{4.3}$$

Substituting for $n_2^2 - n_1^2$ from Equation 4.3 into Equation 4.2,

$$2n_1 \Delta n = 2n_1 \sin b + 2 \sin^2 b$$

Therefore,

$$\Delta n = \sin b + \frac{\sin^2 b}{n_1}$$

Thus the determination of the angle b enables Δn to be found if n_1 is known.

4.2 Critical Angle Methods

In these methods, one surface of a glass block of accurately known refractive index is covered with a parallel-sided slab or film of the material of unknown refractive index. The critical angle is then measured at which light is refracted into the glass block when the angle of incidence at the covered surface is 90° (grazing incidence). For angles of incidence less than 90°, the angle of refraction will be smaller than the critical angle of refraction. Observation of the refracted beam with a telescope will reveal a particular position, depending on the refractive indices of the block and the material of the slab or film, at which the field of view is divided into light and dark portions.

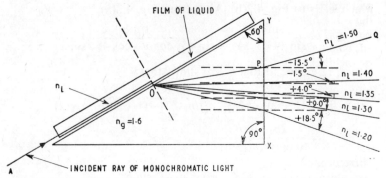

Fig. 4.3. Light incident at 90° on a liquid film covering the hypotenuse face of a right-angled prism

An example appropriate to refractometers which make use of critical angles is considered in Fig. 4.3. A prism of glass of refractive index $n_g = 1.6$ at a wavelength λ (frequently sodium light of wavelength 5,893 Å) has angles of 90°, 60° and 30°, as shown. The hypotenuse face is covered with a film of liquid of refractive index n_l. A ray AO of monochromatic light of wavelength λ is incident at 90° on the film of liquid. The angle of emergence of the light will depend on n_l. Calculation shows that this angle varies between $-15.5°$ for $n_l = 1.5$ and $+18.5°$ for $n_l = 1.2$. For example, if $n_l = 1.5$, at the film–glass interface

$$\frac{\sin I}{\sin R} = \frac{\sin 90°}{\sin C} = \frac{1}{\sin C} = \frac{n_g}{n_l} = \frac{1.6}{1.5}$$

where C is the critical angle of glass for light from a film of refractive index 1.5. Therefore,

$$C = \sin^{-1} 0.937 = 69°33'$$

The angle of incidence of this ray OP at the face XY of Fig. 4.3 is therefore 9°33'; so the angle of emergence of PQ is θ, given by:

$$\frac{\sin \theta}{\sin 9°33'} = 1.6$$

Therefore,

$$\theta = \sin^{-1} 0.267 = 15.5°$$

If the film of liquid is illuminated by a lamp (for example, a sodium lamp providing light of wavelength $\lambda = 5,893$ Å) placed so that AO is the extreme ray but other rays making smaller angles of incidence than 90° are also present, it is clear that PQ

will represent a well-defined boundary between the light and dark regions in the field of the emergent light. Thus, any ray beyond PQ and the normal at P would, if considered to be reversed, be internally reflected at the hypotenuse face because it would be incident at this face at an angle greater than the critical angle. Correspondingly, any ray emergent from the face XY can only have emergent angles numerically less than 15·5° or be on the other side of the normal.

It can be seen that emergent angles between $+18\cdot5°$ and $-15\cdot5°$ (i.e. a range of 34° or 2,040′) correspond to a range of 1·2 to 1·5 in refractive index values. Hence approximately 1′, which is measurable, corresponds to a change of 0·00015 in refractive index.

Wollaston's method of measuring the refractive index n_l of a liquid makes use of a cube $ABCD$ of glass of known refractive index $n_g > n_l$ (n_g can be found, if required, by the 'apparent depth' method and a travelling microscope). Only a small quantity of the liquid is needed, a film of it being sandwiched between the face BC of the glass cube and a flat sheet of black ebonite or other suitable material (Fig. 4.4). This face BC is vertical, so $ABCD$ in Fig. 4.4 is a plan view of the cube. The face AB of the block is illuminated with diffuse light (for example, from a sodium lamp behind a ground glass screen or, if chromatism is tolerated, from an illuminated white card. With the eye at E on the other side of the block, two pins are placed so that they are collinear

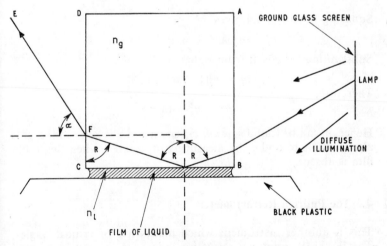

Fig. 4.4. *Wollaston's method for determining the refractive index of a small drop of liquid*

with the dividing line between the light and the dark region in the field of view. The angle α shown in Fig. 4.4 is measured with a protractor, the normal to the face CD and the direction of the ray FE being located by lines drawn on a sheet of paper on a drawing board under the block. A more precise method for measuring the angle α would be to set the block on the table of a spectrometer.

When the angle R in Fig. 4.4 is the critical angle, the emergent boundary ray FE will result. For angles equal to, and exceeding, the critical angle R, reflection occurs at the glass–liquid interface and the emergent light is below FE; for angles less than the critical angle, reflection does not occur, so light does not emerge from the face CD above FE in the diagram. When angle R is the critical angle,

$$\frac{n_g}{n_l} = \frac{\sin 90°}{\sin R} = \frac{1}{\sin R}$$

Therefore,

$$n_l = n_g \sin R \qquad (4.4)$$

Considering the refraction at the surface CD in Fig. 4.4,

$$\frac{n_g}{n_{air}} = \frac{\sin \alpha}{\sin (90° - R)} = \frac{\sin \alpha}{\cos R}$$

Therefore, since $n_{air} = 1$,

$$\cos R = \frac{\sin \alpha}{n_g} \qquad (4.5)$$

Squaring Equation 4.4 gives:

$$n_l^2 = n_g^2 \sin^2 R = n_g^2(1 - \cos^2 R)$$

Substituting for $\cos R$ from Equation 4.5,

$$n_l^2 = n_g^2[1 - (\sin^2 \alpha)/n_g^2]$$

Therefore,

$$n_l^2 = n_g^2 - \sin^2 \alpha$$

Hence, n_l can be found since n_g is known and α is determined. It is also possible to find n_g by measuring the angle α when the liquid film is absent.

4.3 The Pulfrich Refractometer

This is another instrument which makes use of critical angles. The liquid of unknown refractive index n_l is placed in a small glass cell cemented to the top polished face of a glass block of known high refractive index n_g (n_l must be less than n_g). For some liquids, the cell can be omitted and a small drop of the liquid simply formed on the top face. The Pulfrich refractometer can also be used for finding the refractive index of a transparent solid: the solid must have a polished face which can be optically contacted to the top face of the block by means of an intermediate film of a liquid with high refractive index—for example, α-mono-bromonaphthalene ($n = 1.66$) or methylene iodide (for which $n = 1.74$).

Only a certain range of values of n_l, the refractive index of the liquid in the cell, may be measured. As shown in Fig. 4.5(a), the vertical left-hand face of the block, which is towards the incident light, is opaque; the opposite right-hand face, from which the light emerges, is optically polished. A slightly convergent beam of monochromatic light (for example, from a sodium lamp) makes

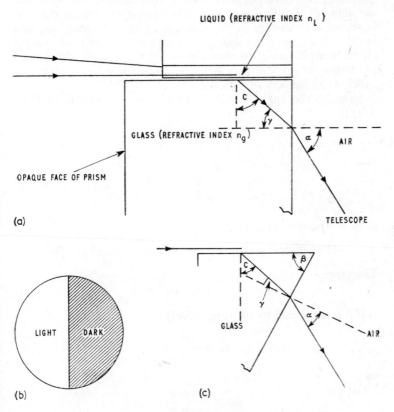

Fig. 4.5. *The Pulfrich refractometer*

glancing incidence with the liquid–glass interface. The light which emerges from the right-hand face in the diagram is viewed with a telescope which can be moved round a mount provided with a vertical divided circle; in this way, the angle α can be measured accurately. This telescope is often fitted with an auto-collimating device so that its position when normal to the face of the block can be located.

Light is refracted into the glass block over the whole length of the cell containing the liquid. The telescope brings all emergent rays in a particular direction to a particular focus. The ray in the incident beam at an angle of incidence of 90° to the liquid–glass interface is refracted into the block at an angle of refraction C, the critical angle. This ray is then incident at an angle γ to the right-hand face of the block and emerges at the angle α. At this angle, the field of view in the observing telescope will be divided into light and dark portions [Fig. 4.5(b)] because this emergent ray defines the critical boundary. The sharp line dividing the field is set on a cross-wire in the eyepiece of the telescope, and the angle α is measured accurately with a vernier against the divided circle. Some illumination comes from rays with angles of incidence at the top face of less than 90°. These rays are refracted into the block at angles less than C.

Usually, the angle β of the block is 90°. If this is so [Fig. 4.5(a)],

$$\frac{n_g}{n_{\mathrm{air}}} = \frac{n_g}{1} = \frac{\sin \alpha}{\sin \gamma}$$

and

$$\frac{n_g}{n_l} = \frac{\sin 90°}{\sin C} = \frac{1}{\cos \gamma}$$

so $n_l = n_g \cos \gamma$. Therefore,

$$n_l^2 = n_g^2 \cos^2 \gamma = n_g^2(1 - \sin^2 \gamma) = n_g^2[1 - (\sin^2 \alpha)/n_g^2]$$

Therefore,

$$n_l^2 = n_g^2 - \sin^2 \alpha \qquad (4.6)$$

If β is not equal to 90° [Fig. 4.5(c)],

$$n_g = \frac{\sin \alpha}{\sin \gamma}$$

so

$$\sin \gamma = \frac{\sin \alpha}{n_g}$$

Also,

$$\frac{n_g}{n_l} = \frac{\sin 90°}{\sin C}$$

so

$$n_l = n_g \sin C$$

But $C = \beta - \gamma$. Therefore,

$$n_l = n_g \sin (\beta - \gamma)$$
$$= n_g(\sin \beta \cos \gamma - \cos \beta \sin \gamma)$$
$$= n_g[(\sin \beta)\sqrt{(1 - \sin^2 \gamma)} - \cos \beta \sin \gamma]$$
$$= n_g \left[(\sin \beta) \sqrt{\left(\frac{n_g^2 - \sin^2 \alpha}{n_g^2}\right)} - \frac{\cos \beta \sin \alpha}{n_g}\right]$$

Therefore,

$$n_l = (\sin \beta)\sqrt{(n_g^2 - \sin^2 \alpha)} - \sin \alpha \cos \beta \qquad (4.7)$$

which reduces to Equation 4.6 when $\beta = 90°$. When $\beta < 90°$, α may be positive or negative.

The Pulfrich refractometer is kept at a constant temperature because dn/dT (the variation of refractive index with temperature) is about 1×10^{-4} per degree Celsius. This means that the accuracy of which some instruments are capable (2 parts in 10^5) is nullified unless the temperature is known and maintained to within ± 0.1 deg C.

An accuracy of 0.00002 in the determination of a refractive index in the range 1.3 to 1.95 is claimed for some instruments.

4.4 The Abbe Refractometer

The Abbe refractometer is also a critical angle instrument: a drop of the liquid of unknown refractive index is sandwiched as a film between the polished hypotenuse face of a refracting prism of high refractive index glass and the ground (roughened) hypotenuse face of a second 'illuminating' prism (Fig. 4.6). This 'sandwiching' of the liquid prevents any evaporation of the film; such evaporation would cause cooling and change the refractive index. Monochromatic light is directed from a mirror into the lower polished surface of the illuminating prism. Some of this light is scattered by the ground hypotenuse face at grazing incidence (angle of incidence 90°) to the liquid–prism boundary (i.e. at the liquid–prism boundary, $\sin I = 1$).

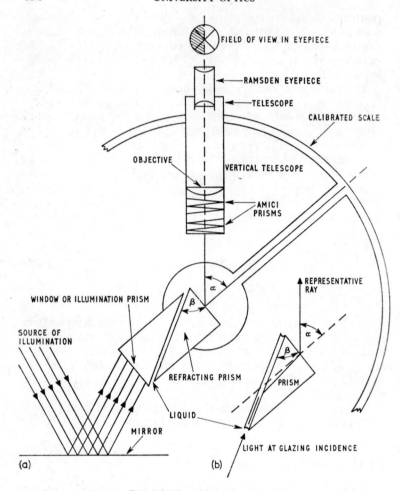

Fig. 4.6. The Abbe refractometer

Fig. 4.6(b) shows the representative ray defining the boundary between the light and dark regions of the field of view. If the angle of the prism shown is β, the analysis is as in Section 4.3 where $\beta < 90°$. Hence, Equation 4.7 applies and, since α is negative in the arrangement shown in Fig. 4.6(b), the equation becomes:

$$n_l = (\sin \beta) \sqrt{(n_g^2 - \sin^2 \alpha)} + \sin \alpha \cos \beta \qquad (4.8)$$

The viewing telescope is fixed in a vertical position but the calibrated arc-shaped scale moves with an attached circular platform on which the prisms are mounted. The goniometer scale, which is read by another eyepiece, can be calibrated in degrees or in direct readings of the refractive index of the solution between the two prisms. Other features of this instrument are shown in Fig. 4.6(a).

The performance of the Abbe refractometer is outstanding in that the refractive index can be measured to the fourth decimal place provided the temperature is maintained constant to within $\pm 0 \cdot 1$ deg C, only a few drops of the liquid are needed, and the determination can be made in less than 5 min.

If white light illumination is used, dispersion giving chromatism results in an ill-defined boundary between the light and dark regions of the field of view. Amici prisms near the telescope objective ensure that the refractive index may be measured for light of wavelength 5,893 Å. As shown in Fig. 4.6(a), they consist of two direct vision prisms which can be rotated by equal and opposite amounts about the telescope axis to vary the dispersion of the combination. Rays emerging from the refracting prisms may be dispersed owing to variation of refractive index with wavelength. The Amici prisms recombine these rays so that the critical boundary line is achromatised. The rotation of the prisms from their position when the instrument is used with light of wavelength 5,893 Å gives a measure of the dispersion ($n_C - n_F$) of the liquid sample (see Equation 2.14).

To check the Abbe refractometer, the illuminating prism may be removed and replaced by a small block of glass of known refractive index with a small drop of α-monobromonaphthalene on the face of the refracting prism. The instrument is tilted to prevent this block from sliding off.

4.5 The Hilger–Chance Refractometer

In the critical angle type refractometers described, the eyepiece cross-wires are set on a demarcation line or boundary between light and dark areas of the field of view. This boundary has to be well defined for accuracy in measurement. Furthermore, the refractive index of the glass must be greater than that of the liquid specimen, and the value of refractive index determined is for a thin film of the liquid.

The Hilger–Chance refractometer requires about 2 cm³ of liquid—which is recoverable—placed within the V-section channel between the optically polished hypotenuse faces of two 45° right-angled prisms of glass of any refractive index. A collimated beam of monochromatic light is incident normally, as shown in Fig.

4.7(a). The angle at which this beam emerges is measured by a telescope movable over a divided circle. The Ramsden eyepiece of the telescope has two short etched parallel lines on a flat glass disc in its focal plane, and the setting is made on a fine line which is the image of a fine line ruled on a flat glass over the entrance slit. The image is set exactly midway between the parallel lines in the eyepiece [Fig. 4.7(b)].

The refractive index can be determined from the angle of inclination of the telescope to the direction of the incident light by use of Equation 4.9. The scale is not calibrated directly in readings of refractive index because this instrument may be used to measure the variation of refractive index with wavelength of light. Filters placed over the entrance slit in conjunction with

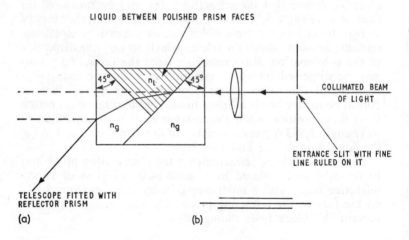

Fig. 4.7. The Hilger–Chance refractometer

various discharge lamps (cadmium, sodium and cadmium–mercury discharge lamps) may be used to isolate at least 10 known wavelengths. The constants A and B can be found in Cauchy's dispersion equation:

$$n = A + \frac{B}{\lambda^2}$$

Referring to Fig. 4.8, Snell's law is applied at the three surfaces where the incident ray is deviated. At the first deviation:

$$\frac{n_g}{n_l} = \frac{\sin (90° - \theta)}{\sin 45°} = (\cos \theta)\sqrt{2}$$

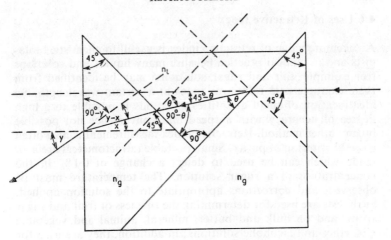

Fig. 4.8. Ray-trace through the prisms of a Hilger–Chance refractometer

So

$$\left(\frac{n_g}{n_l}\right)^2 = 2\cos^2\theta$$

At the second deviation:

$$\frac{n_l}{n_g} = \frac{\sin(45° - x)}{\sin\theta}$$

Therefore,

$$\left(\frac{n_l}{n_g}\right)^2 = \frac{\sin^2(45° - x)}{1 - \cos^2\theta} = \frac{\sin^2(45° - x)}{1 - n_g^2/2n_l^2} = \frac{2n_l^2\sin^2(45° - x)}{2n_l^2 - n_g^2}$$

Therefore,

$$\frac{\sqrt{(2n_l^2 - n_g^2)}}{n_g\sqrt2} = \frac{\cos x}{\sqrt2} - \frac{\sin x}{\sqrt2}$$

Substituting for $\cos x$ and $\sin x$ from $n_g = (\sin y)/(\sin x)$ at the third deviation:

$$\frac{\sqrt{(2n_l^2 - n_g^2)}}{n_g} = \frac{\sqrt{(n_g^2 - \sin^2 y)}}{n_g} - \frac{\sin y}{n_g}$$

Therefore,

$$2n_l^2 - n_g^2 = n_g^2 - \sin^2 y + \sin^2 y - 2(\sin y)\sqrt{(n_g^2 - \sin^2 y)}$$

or

$$n_l^2 = n_g^2 - (\sin y)\sqrt{(n_g^2 - \sin^2 y)} \tag{4.9}$$

As n_g is known and y is determined, n_l is found.

4.6 Uses of Refractive Index

An accurate value of refractive index is useful to chemists, biologists and in clinical practice because many liquids and solutions (for example, salt and sugar solutions) may be identified from their characteristic refractive index at a certain wavelength. So identification of small quantities of liquids is possible to a high degree of accuracy with a speed and convenience not possible by any other method. Refractive index measurement is also often a useful criterion of purity. Small portable refractometers are now made which can be used to detect a change of 0.1% in the concentration of a sugar solution. The temperature must be observed, and corrections appropriate to the solution applied. Such tests are used for determining the ripeness of fruit and sugar crops, and on milk and butter, mineral, animal and vegetable oils, ether and alcoholic solutions; in addition, they are used for determining the proportions of the components of many binary mixtures.

Theoretical use of refractive index measurement is made in molecular refraction studies. Molecular refraction is an additive and constituent property used in a manner analogous to Sugden's parachor involving surface tension, molecular rotation in the rotation of plane polarised light and molecular susceptibility involving magnetic susceptibility measurements.

In 1882, Lorentz and Lorenz derived an equation for the molecular refraction R as

$$R = \frac{n^2-1}{n^2+2} \frac{M}{\varrho}$$

where M is the molecular weight, ϱ is the density, and n is the refractive index of the substance. As n varies with the wavelength of light, values of R are obtained for various wavelengths. For a wavelength of $5{,}893\,\text{Å}$, values of R for some elements, radicals and structural features are: carbon, 2.148; hydrogen, 1.100; chlorine, 5.967; a double bond, 1.733; a triple bond, 2.398; oxygen in CO, 2.211; oxygen in ethers, 1.643; oxygen in OH, 1.525; CH_2 in homologous series for esters 4.606, and for alcohols 4.634.

It follows that the molecular refraction R for chloroform ($CHCl_3$) is:

$$2.418 + 1.100 + 3(5.967) = 21.419$$

For chloroform, the experimentally determined values are $n = 1.45$, $M = 119.4$ and $\varrho = 1.50$. Substituting in the Lorentz–

Lorenz equation:

$$R = \frac{(1 \cdot 45)^2 - 1}{(1 \cdot 45)^2 + 2} \times \frac{119 \cdot 4}{1 \cdot 50} = 21 \cdot 5$$

The theoretical and measured values of R are seen to agree quite well. In practical work, the student may select suitable organic liquids and determine the molecular refraction of various groups: for example, R for CH_2 from measurements on n, M and ϱ for methyl alcohol (CH_3OH) and ethyl alcohol (C_2H_5OH).

Exercise 4

1. Outline, very briefly, the methods available for the accurate measurement of the refractive index of a liquid, and describe one of them.
   A hemisphere of radius r made of glass of refractive index μ is supported with its plane face horizontal and facing downwards. A small object on an illuminated horizontal stage is at a distance r/μ below the centre of the face, and the space between is filled with a liquid of the same refractive index μ. Show that there will be an exact virtual image at a distance $r\mu$ below the centre, however wide the angle of the beam used. What is the practical application of this arrangement? (L. Anc.)

2. Describe and explain the use of a precision instrument for determining the refractive index of liquids. A 60° glass prism has a refractive index of 1·6000 and 1·5852 respectively for the blue and red hydrogen lines, and is set to transmit the red line at minimum deviation. Calculate (a) the angular separation of the red and blue beams on emergence from the prism, and (b) the linear separation of the lines in the spectrum formed by a lens of mean focal length 50 cm, the light incident on the prism being a parallel beam. (L. Anc.)

3. Give the theory of a critical angle method for determining with precision the refractive index of a liquid using a spectrometer and a prism of material of known refractive index. Outline the practical details and discuss the relative merits of this method and those which do not depend on the measurement of critical angle. What would be the effect in this method on the appearance of the limiting boundary in the field of view if a mercury lamp were substituted for a monochromatic source? (L. G.)

4. Describe a critical angle precision refractometer for liquids. Give the essential theory.
   A rectangular block of glass stands on a ground glass screen, illuminated from below, and there is a film of water of refractive index 1·33 between the glass surfaces. Find the least refractive index of the glass block if any light from the base is to emerge through a vertical face. (L. Anc.)

5. Describe and explain how to determine experimentally the dispersive power of a transparent liquid, using a critical angle method for measuring any refractive index involved. (L. Anc.)

6. Describe a method of measuring with high precision the refractive index of a liquid.

A cell of square cross section has a diagonal partition across it, the walls of the cell and the partition being made of optically flat glass sheet. One half of the cell contains water and the other half contains a transparent aqueous solution. A parallel beam of light enters one face of the cell parallel to the normal, and its angular deviation is measured by means of a telescope which is capable of resolving two objects not less than 2 cm apart at 1 km distance. Calculate the smallest difference of refractive index between water and a solution that could be detected in this way. (L. Anc.)

7. Give the theory of a critical angle method of determining the refractive index of a liquid by the use of a glass prism of known refracting angle and refractive index.

A cube of glass of refractive index 1·624 is used in a critical angle experiment to measure the refractive index of a liquid. Between what limits must the refractive index lie if it is to be measurable with this apparatus? (L. Anc.)

8. Give an account of a precise laboratory method which does not rely on the principles of interferometry, for measuring the refractive index of liquids. It is to be assumed that the liquids are only available in small quantities. (L. Anc.)

9. Discuss the relative merits of critical-angle and non-critical-angle methods in the precision refractometry of liquids.

Describe in detail a method of investigating the variation, with concentration, of the refractive index of a *dilute* solution. (L. P.)

10. Describe an instrument which can be used for highly accurate determinations of the refractive indices of liquids.

A glass cell with a square base has a glass partition which forms one diagonal of the cell. Show how this cell could be used to find the difference of refractive index between a dilute aqueous solution in one half of the cell and pure water placed in the other half. (L. P.)

11. Describe and explain an experiment using a spectrometer to determine the refractive index of glass in the form of a prism. Derive the equation used to calculate the refractive index.

An equi-convex lens has surfaces of radii of curvature 17·1 cm and a focal length of 15·0 cm for sodium light. Calculate the angle of minimum deviation for a 60° prism made from the same glass when used with sodium light. (L. G.)

CHAPTER 5

The Velocity of Light

5.1 Electromagnetic Radiation

In 1873, Maxwell published his electromagnetic theory, the fundamentals of which are considered in Chapter 1, Volume 2. This theory deals with the fact that an electrostatic field which is varying is accompanied by a varying magnetic field, and vice versa. Thus, the effect of accelerating electric charge, as when a varying current passes through a conductor, is to produce an electromagnetic field in the surrounding medium. Maxwell established the differential equations relating the variation in space of the electrostatic field to the rate of variation with time of the magnetic field and the corollary, the space rate of variation of the magnetic field to the time variation of the electric field. He further showed that these variations are propagated as a wave motion with a speed v given by

$$v = \frac{1}{\sqrt{(\mu\varepsilon)}} \tag{5.1}$$

where μ is the magnetic permeability of the medium and ε its permittivity (dielectric constant).

Equation 5.1 indicates that the velocity of propagation in free space *(in vacuo)* of an electromagnetic wave is $1/\sqrt{(\mu_0\varepsilon_0)}$, where μ_0 is the magnetic permeability and ε_0 the permittivity of free space. In centimetre–gramme–second (c.g.s.) electrostatic units (e.s.u.), ε_0 is unity. In these same units, $\mu_0 = 1/r^2$, where

$$r = \frac{\text{electromagnetic unit (e.m.u.) of charge}}{\text{electrostatic unit of charge}}$$

and r can be found by experiment. It follows that the velocity of electromagnetic radiation in free space is:

$$\frac{1}{\sqrt{(\mu_0 \varepsilon_0)}} = \frac{1}{\sqrt{(1/r^2)}} = r \text{ cm s}^{-1}$$

The important fact arises that this ratio r found by experiment is numerically equal to the velocity of light c in free space as determined before Maxwell's theory by Foucault in 1850 (Section 5.4) and subsequently by Michelson in 1926 (Section 5.5) and Bergstrand in 1950 (Section 5.3).

Maxwell concluded that visible light is a form of electromagnetic radiation comprising varying electric and magnetic field vectors shown to be perpendicular to the direction of propagation: i.e. a transverse wave is concerned. In 1887, Hertz generated electrically electromagnetic waves of a few metres in length ('radio' waves) which were predicted by Maxwell.

It is now fully established that the electromagnetic waves, which comprise the *electromagnetic spectrum*, have a wide range of wavelengths: from radio waves of several thousand metres in length, to gamma radiation of wavelengths 10^{-11} cm and less (Fig. 5.1). Within this extremely wide spectrum, visible light (i.e. electromagnetic radiation to which the eye is sensitive) occupies only a very narrow region extending roughly over an octave from 3,800 Å to 7,600 Å approximately.

All the forms of electromagnetic radiation are propagated with the same velocity c in free space, where

$$c = (2 \cdot 99793 \pm 0 \cdot 000003) \times 10^{10} \text{ cm s}^{-1}$$

$$= 3 \times 10^{10} \text{ cm s}^{-1} \text{ approx.}$$

As $f = c/\lambda$, where f is the frequency and λ the wavelength, the frequency values given in Fig. 5.1 may be calculated from the various wavelengths. In the visible spectrum, violet light of wavelength 4,000 Å will have a corresponding frequency of $(3 \times 10^{10})/(4 \times 10^{-5}) = 7 \cdot 5 \times 10^{14}$ Hz; similarly, red light of wavelength 7,000 Å will have a frequency of $4 \cdot 3 \times 10^{14}$ Hz.

The velocity of electromagnetic radiation in free space or the velocity of light may therefore be determined experimentally by the use of visible light or by the use of other radiations, in particular radio waves; this velocity can also be determined by the measurement of the ratio of the e.m.u. of charge to the e.s.u. of charge. In general, accurate methods undertaken before World War II utilised visible light; since the War, the use of short-wavelength radio waves has become prominent. The measurement of the ratio of the units of electric charge in the two systems

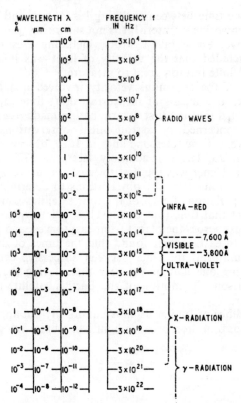

Fig. 5.1. The electromagnetic spectrum: 1 micron (μm) $= 10^{-6}m = 10^{-4}cm$;
1 Ångstrom (Å) $= 10^{-8}cm$; 1 hertz (Hz) $= 1$ cycle per second (c/s)

of units is the concern of electricity and is really a verification of
Maxwell's concepts and not a fundamental method of measuring
velocity. It will not be described here.

5.2 Outlines of the Early Astronomical Methods of Determining the Velocity of Light

Following the first measurement of the velocity of sound by
Mersenne (1588–1648) who timed echoes (his result was 1,038
ft s^{-1}), attempts were made to measure the velocity of light by
similar methods. Galileo (1564–1632) tried with the aid of a co-
worker, a lantern and a pail: the co-worker, on receiving a flash
of light from Galileo's lantern, exposed his lantern, and Galileo

recorded the time between exposing his lantern and seeing light from his co-worker's lantern! It is not surprising that the attempt was abortive. Some investigators, including Descartes (1596–1650), concluded that the velocity of light was infinite, others that it was finite but too great to measure.

In view of the enormous velocity involved and the lack, in early work, of a means of measuring short time intervals, it is to be expected that the first determination made use of the great distances concerned in astronomical observations. In 1608, Galileo invented the telescope and, in 1610, observed with it the satellites of Jupiter. Considerably later, in 1676, the Danish astronomer Römer was the first to measure the velocity of light by making a long series of observations to determine the mean period of revolution of each of the four satellites which could be discerned at that time (out of the 11 now known to exist). The satellites pass across the planet from east to west and pass behind from west to east. Römer found that the times of their eclipses were not uniform, the time between successive eclipses being longer during one half of the year than during the other half. By comparison, the motion of the Earth's satellite (the Moon) was regular.

The satellites of Jupiter do move uniformly but appear to have a non-uniform motion because light has a finite velocity and the

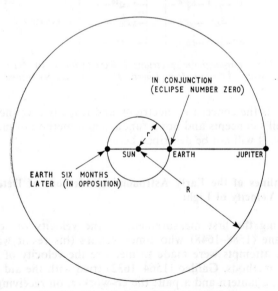

Fig. 5.2. The principle of Römer's method

distance between Jupiter and the Earth varies in the course of the year by the diameter of the Earth's orbit. This diameter was determined in about 1675 but with insufficient accuracy. Even now, terrestrial methods are preferred to astronomical ones because of uncertainty of the size of the Earth's orbit. Fig. 5.2 illustrates the principle of Römer's method. The eclipse number zero is seen on Earth $(R-r)/c$ after the conjunction occurs. If T is the time of revolution of a satellite about Jupiter after a time nT between conjunction and opposition, the nth eclipse occurs. This is seen $(R+r)/c$ later on Earth. Let T_1 be the time for the first n eclipses, and T_2 the time for the second successive sequence of n eclipses; then

$$nT = T_1 + \frac{R-r}{c} - \frac{R+r}{c} = T_1 - \frac{2r}{c}$$

and

$$nT = T_2 - \frac{R-r}{c} + \frac{R+r}{c} = T_2 + \frac{2r}{c}$$

Therefore,

$$T_1 - \frac{2r}{c} = T_2 + \frac{2r}{c}$$

or

$$T_1 - T_2 = \frac{4r}{c}$$

$T_1 - T_2$ was found to be about 2,000 s, and r is known to be 93×10^6 miles, giving $c = 186,000$ miles s^{-1}. Römer gave the result $c = 3 \cdot 1 \times 10^{10}$ cm s^{-1}, an outstanding achievement in such early work.

In 1728, the English astronomer Bradley measured the velocity of light by observing the *aberration of light* from stars. He is reputed to have thought of this method after observing apparent changes of wind direction as his own speed and direction changed while he was sailing.

A star observed from the Earth moving in its orbit will not appear to stay in exactly the same place. To a very small degree, this effect is due to parallax whereby nearer stars to the Earth appear to move against the background of more distant stars. The greater part of this apparent motion is due to aberration, which is the change in apparent direction caused by the movement of the observer. Fig. 5.3(a) illustrates that, in order to receive light from a star, a telescope must be tilted at an angle α to the true direction. Six months later, the angle of aberration will be $-\alpha$. Hence, two observations made at an interval of six months will give the constant, or angle, of aberration. Bradley made observations on the star γ Draconis, declination $51° 29.7'$, which is overhead at Greenwich on attaining maximum altitude. In fact, the telescope was not tilted as shown but was secured to a chimney at Greenwich; α was measured by the movement of the image of the star on the eyepiece cross-wires.

If a modern value of about 93×10^6 miles is taken for the distance between the Earth and the Sun, the Earth's velocity v is 18.6 miles s^{-1} (the rotation of the Earth alters this velocity by about 0.18 miles s^{-1} at Greenwich). The aberration constant α

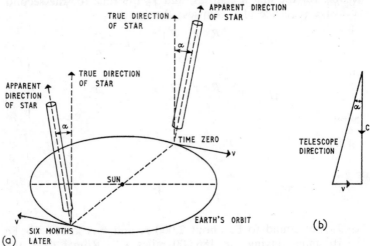

Fig. 5.3. Principle of Bradley's method of observing the aberration of light from a star

is about $20''$. Parallax effects are about $0.02''$. From Fig. 5.3(b) it can be seen that $\tan \alpha = v/c$ when the angle of elevation of the star at maximum altitude is $90°$. Hence,

$$c = \frac{18.6}{\tan 20''}$$

As $\tan \alpha = \alpha$,

$$\tan 20'' = \frac{20 \times 2\pi}{360 \times 3{,}600}$$

Therefore,

$$c = \frac{18.6 \times 1{,}296 \times 10^3}{40\pi} = 190{,}000 \text{ miles } s^{-1} \text{ approx.}$$

Bradley's result, taking into account corrections not given in this simple calculation, was $3 \cdot 083 \times 10^{10}$ cm s^{-1}. This agrees well with Römer's result; Bradley and Römer used the same value for the diameter of the Earth's orbit.

5.3 Optical Shutter Methods for Determining the Velocity of Light

In view of the short times involved for the passage of light over distances of a few miles in a terrestrial method (the time is approximately $3 \cdot 3$ μs for a distance of 1 km) a procedure involving short flashes of light separated by regular short intervals of time (i.e. the use of a rapidly opening and closing shutter) is a promising approach. This was undertaken in the *toothed wheel* experiment of Fizeau in 1849, the first terrestrial method of measuring the velocity of light (Fig. 5.4).

Light from a slit S is focused by the achromatic lens L_1 at a point P near the circumference of a toothed wheel W after reflection by a transparent glass plate M_1. P is in the focal plane of lens L_2 so that a collimated beam travels to lens L_3, which is at the centre of curvature of the mirror M_2. After reflection at M_2, the light retraces its path back via L_3 and L_2 and passes P to the eyepiece L_4. The eye at E, therefore, sees an image of the source of light.

The wheel W, which has N equally spaced, bevelled and blackened teeth round its circumference, is rotated. The light is thus

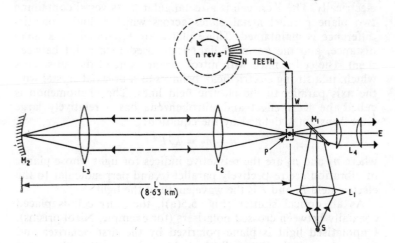

Fig. 5.4. Fizeau's method

alternately allowed to pass through a gap between the teeth or it is stopped by a tooth at P. The wheel is rotated at gradually increasing speed until no light is visible at E. This will occur at several speeds: suppose the lowest one is n rev s^{-1}. This means that light reaching M_2 and being reflected cannot pass to E because a tooth has moved to the position of the preceding gap. The speed of the rotating wheel is determined stroboscopically relative to standard tuning forks.

Let L be the distance from P to M_2. Then, at the minimum speed of rotation when no light is seen at E, the light travels a distance $2L$ in the time T taken for one tooth to move to the position of the adjacent gap; clearly $T = 1/2Nn$. Hence, the velocity of light c is given by

$$c = \frac{2L}{1/2Nn} = 4LNn$$

In Fizeau's experiment, $L = 8.63$ km and $N = 720$. In 1874, Cornu repeated this method with a path length L of 22·91 km, N was either 150 or 200, and the maximum speed of the wheel was 900 rev s^{-1}. Perrotin made a new determination in 1900 using a base line of 40 km, and he corrected the determined velocity because the value *in vacuo* is 1·00028 times the velocity in air. He obtained $c = (299,860 \pm 80)$ km s^{-1}.

In 1926, Karolus and Mittelstaedt devised a method essentially similar to that of Fizeau. They replaced the toothed wheel by a *Kerr cell*, which, placed between crossed polarisation filters, provides a very convenient optical shutter which can be operated electrically. The Kerr cell is a rectangular glass vessel containing two plane parallel metal plates across which a high potential difference is maintained. These plates are separated by a small distance, and the beam of light is focused to a point between them. The cell is filled with nitrobenzene—one of the substances which, in a strong electric field, behaves as a uniaxial crystal with the axis parallel to the electric field lines. The phenomenon is called the Kerr effect and nitrobenzene has a relatively large Kerr constant k defined by the equation:

$$n_1 - n_2 = k\lambda E^2$$

where n_1 and n_2 are the refractive indices for light whose planes of vibration are respectively parallel to and perpendicular to the electric field E, and λ is the wavelength of the light.

As an optical shutter [Fig. 5.5(a)], the Kerr cell is placed coaxially between crossed polarisers (for example, Nicol prisms). Unpolarised light is plane polarised by the first polariser and passes through the Kerr cell (the light is convergent to the centre

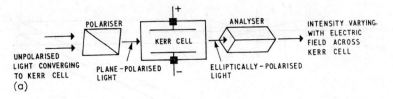

(a)

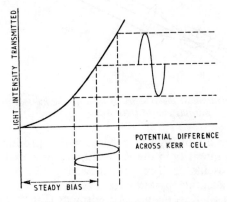

(b)

Fig. 5.5. The use of a Kerr call as an optical shutter

of this cell); it then passes to the second polariser (analyser), which is rotated so that no light emerges when the potential difference (p.d.) across the Kerr cell is zero. The first polariser is set so that the plane of vibration of the light is at 45° to the direction of the electric field established when a p.d. is applied to the Kerr cell. It is convenient to say that light is transmitted when the p.d. producing the electric field on the Kerr cell is on, and is not transmitted when the field is off; also, that the intensity of light transmitted is proportional to the square of the electric field. After passing through the Kerr cell, the light is elliptically polarised, with a phase difference δ between the ordinary and extraordinary components which is proportional to the square of the applied field. The light intensity I transmitted by the analyser is given by

$$I = I_0 \sin^2 \tfrac{1}{2}\delta$$

where I_0 is the incident intensity.

If an alternating p.d. is maintained across the plates, the intensity of the light transmitted will vary. To ensure that the more linear part of the Kerr cell characteristic (intensity against p.d.) is involved, the alternating p.d. is put in series with a steady p.d. across the cell [Fig. 5.5(b)]. The light intensity then varies at the frequency of the alternating p.d. applied.

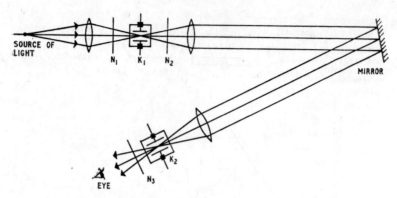

Fig. 5.6. Determination of the velocity of light by the method of
Karolus and Mittelstaedt

Suitable periodic p.d.s of high frequency, which produce periodic fields of high intensity in the Kerr cell, are provided by a valve or transistor oscillator. The frequency may be controlled to a high degree of accuracy by a quartz crystal.

In the experiment of Karolus and Mittelstaedt (Fig. 5.6), N_1, K_1 and N_2 are the Nicol prism polariser, Kerr cell and Nicol prism analyser of the electro-optical shutter associated with the source of light; N_2 is orientated so that the plane of vibration of the light transmitted is at $90°$ to that transmitted by N_1 and the Nicol prism N_3. Light from the source is only transmitted if there is an electric field across K_1, and light is transmitted by N_3 only if there is a field across the Kerr cell K_2 when the light reaches it. K_1 and K_2 are connected across the same source of high-frequency alternating electromotive force (e.m.f.) so that the electric fields across them are in phase. When the frequency of the applied field is such that the time of transit of a light pulse between the two Kerr cells is one-quarter of a period of the alternating field, a light intensity maximum will leave K_1 at maximum field, but by the time it reaches K_2 the field will be zero so that minimum light intensity will be transmitted by N_3.

As the frequency of the electric field is increased, minima of light intensity are observed when the time of transit of the light is $(2n+1)/4$ of a period, where n is an integer. Typically, observations are made for $n = 1, 2, 3, \ldots, 8$. The light path was 300 m, though one determination was with a distance of 41 m, and an oscillator frequency of 3–7 Hz. The velocity of light c was found to be $(299,778 \pm 20)$ km s^{-1}.

Further measurements were undertaken in 1937 and 1941 by Anderson, who used a photoelectric cell to measure the finally

emergent light intensity instead of the optical method used by
Karolus and Mittelstaedt. He made nearly, 3,000 observations
and gave the value of c as $(299,766 \pm 14)$ km s^{-1}.

During 1949 and 1950, Bergstrand undertook a series of determinations with apparatus which was a considerable technical
development on that of Anderson, though similar features were
used. The principle of this method is illustrated by Fig. 5.7. S is
the source of light, the intensity of which is modulated by the
high-frequency oscillator O. This light passes in a collimated beam
from the concave mirror M_1 (S is in effect at the focus of M_1) to
the distant plane mirror M_2 and is returned to a photoelectric
cell P (in effect at the focus of concave mirror M_3). The p.d.
across this photocell P is also provided by the oscillator O.

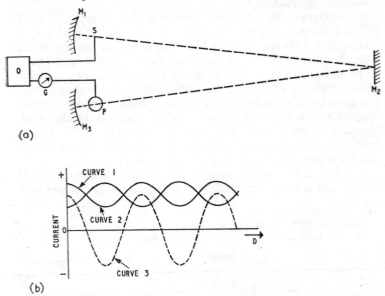

Fig. 5.7. Principle of Bergstrand's method

The sensitivity of the photocell and the intensity of the light
source thus vary at the same high frequency. If the maximum of
the light intensity and the maximum sensitivity of the photocell
occur simultaneously, the light intensity maximum will arrive at
P at a time of high or low sensitivity, depending on the time of
transit of the light over the distance $D = SM_2P$. Hence, the
current from the photocell will vary with D as shown in curve
1 of Fig. 5.7(b).

To enhance the determination of the maxima and minima in curve 1, it is arranged that the curve 2 (where the maximum of the source intensity coincides with the minimum of the photocell sensitivity) is also obtained. It is further arranged that the two currents corresponding to curves 1 and 2 are passed in opposite directions through the measuring instrument G so that the difference curve 3 is obtained with sharply defined positions where the current is zero for specific values of D. To obtain these two curves and the difference between them, the phase of the modulation of the light intensity is reversed periodically by reversing the Kerr cell bias voltage relative to that of the detector; the period of phase reversal is very long compared with that of the light modulation but short compared with the response time of the measuring instrument G.

The apparatus used by Bergstrand (Fig. 5.8) employed a 30 W filament lamp as the source of light S, polaroids 1 and 2 with Kerr cell K between them as the electro-optical shutter, and a concave spherical mirror M_1 (diameter, 46 cm; focal length, 75 cm) to produce a collimated beam to the plane mirror M_3; a plano-convex lens M_2, silvered on the back surface, was used to correct the spherical aberration of M_1 (cf. the Schmidt corrector plate, Section 3.4). A yellow filter was used to minimise the chromatic aberration. From M_3, the light is returned to a similar receiving arrangement of mirrors M_4 and lens-mirror M_5 to reach the cathode of a photomultiplier tube P.

A p.d. of 2 kV root-mean-square (r.m.s.) at a frequency of 8·33 MHz from a crystal-controlled oscillator is applied across both

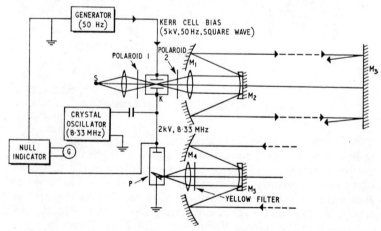

Fig. 5.8. Optical and electrical arrangement of Bergstrand's method

the Kerr cell K and the photomultiplier tube P. In series with this supply to K is a 50 Hz generator providing a peak voltage of 5 kV, ideally of square waveform but in practice simulated by a sine wave of which the top and bottom sections are eliminated. This arranges the comparatively low period reversal of the bias on the Kerr cell.

In the determination of the velocity of light, the distant plane mirror M_3 is moved by a distance d, so the change of the light path is $2d$. This is preferable to relying on the measurement of the total distance D; with this latter method, errors may arise owing to time lag between light arriving at the photomultiplier tube and the emission of electrons to give a signal in the null detector.

If the frequency of the high-frequency oscillator is f (actually 8·33 MHz), the time between successive maxima of light intensity is $1/f$. The null detector indicates two null points for each phase difference of 2π between the p.d. across the photomultiplier and the optical shutter light output. If the change of the light path corresponding to the separation between neighbouring null points is $2d$, the velocity of light c is given by the equation:

$$c = \frac{2d}{f/2} = 4\,df$$

Since c is known to be about 3×10^8 m s^{-1} and f is 8·33 MHz, the distance d between successive positions of zero output from the null detector is about 9 m.

Bergstrand's first value for c (the group velocity *in vacuo*) was $(299{,}793 \pm 2)$ km s^{-1}, the error being 6 parts per million (6 p.p.m.). Later determinations gave $(299{,}793{·}1 \pm 0{·}3)$ km s^{-1}, an estimated error of 1 p.p.m. In 1956, Edge used this method with some refinements and obtained a value of $(299{,}792{·}4 \pm 0{·}1)$ km s^{-1}, an estimated error of 0·3 p.p.m.

It is noteworthy that Bergstrand's value for c exceeded by about 20 km s^{-1} the weighted mean of the various results obtained by experiments undertaken before 1940, which was given by Dorsey in 1944 as $(299{,}773 \pm 10)$ km s^{-1}; i.e. the 1950 value is 10 km s^{-1} greater than the maximum probable error of the pre-war results. Bergstrand's value agrees, however, with post-war methods using radio waves (Section 5.6). Indeed, one of the reasons for Bergstrand's work was to ensure that there was no difference between the value of c found by radio methods using wavelengths of several centimetres and that found by visible-light methods at wavelengths of about 5×10^{-5} cm. There is consequently no reason to suspect any small variations in the value of c

with the wavelength of the radiation. Despite the refinement and elaboration of the pre-war measurements, unknown sources of small systematic errors seem to have been overlooked.

5.4 Rotating Mirror Methods for Determining the Velocity of Light

Following the pioneer terrestrial method of Fizeau (Section 5.3), Foucault undertook a year later (in 1850) the second terrestrial experiment. The main disadvantage of Fizeau's experiments was the long path length over which the light had to travel. This introduced effects due to variations in the atmospheric conditions which were difficult to estimate. Furthermore, the measurement of the velocity of light in optically dense materials is of great interest as it enables tests to be made of the validity of wave theory and the prediction that $c/v = n$, where c is the velocity of light in free space, and v is the velocity of light in a transparent material of refractive index n relative to that of free space (taken to be unity).

Foucault's method made use of a plane mirror rotated at a

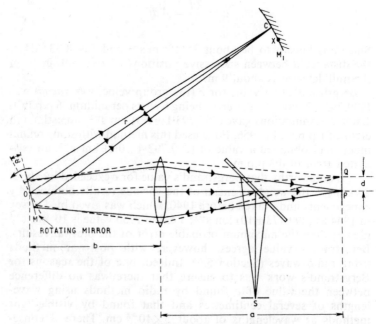

Fig. 5.9. The rotating mirror method of Foucault

measured uniform angular velocity. It leads to an experiment which can be undertaken in the laboratory: both Foucault and Fizeau worked on this method independently.

In Fig. 5.9, the source of light is an illuminated vertical slit S. After reflection from the optically flat glass plate at A, fixed in position at $45°$ to the optical axis of lens L, the light is converged by the achromatic lens L to a focus C at the concave mirror M_1 provided that the front-surface plane mirror M is suitably positioned. M can be rotated at uniform angular velocity about the central vertical axis through O. The centre of curvature of the concave mirror M_1 is at O. If M is stationary, light reflected at M_1 is returned via M and the lens L to traverse the plate A and form an image of the slit S at P on a screen which is viewed by a travelling microscope.

When the mirror M is rotated about the axis through O, the image at the screen will be displaced from P to Q through a distance d. This displacement is due to the fact that the mirror M will rotate through an angle α during the time that the light takes to travel from M to C and back to M. With the rapid rotation of M necessary, the intermittent image formed at P will be repeated so rapidly that it will appear to be continuous (because of the persistence of vision).

Let $OC = r$, $LO = b$ and $LP = a$, where r is large compared with a and b. $AP = AS$ because A is a point on the plane sheet of glass. Rotation of the mirror M through an angle α causes the reflected ray to be deflected through an angle 2α.

Whilst the mirror M is turning through the angle α, the light travels a distance $2r$ to C and back and takes a time $T = 2r/c$, where c is the velocity of light. If the mirror is rotating uniformly at n rev s^{-1}, the time it takes to rotate through the angle α is $\alpha/2\pi n$ (where α is in radians). Therefore,

$$T = \frac{2r}{c} = \frac{\alpha}{2\pi n}$$

Therefore,

$$c = \frac{4\pi n r}{\alpha} \qquad (5.2)$$

Both n and r are known, and α can be found if d is measured. To express α in terms of d, consider that the image of the slit is displaced a distance d from P to Q by rotation of the mirror M through the angle α. This could occur if the light from the mirror M_1 came from X instead of C. As a mirror rotation of α deflects a light ray through 2α, clearly $XC = 2\alpha r$. The rotation of the mirror M only alters to a small extent the direction of the light;

the distances are not changed. For the convex lens L,

$$\frac{\text{length of image}}{\text{length of object}} = \frac{\text{distance from image to centre of lens}}{\text{distance from object to centre of lens}}$$

Therefore,

$$\frac{d}{2\alpha r} = \frac{a}{r+b}$$

or

$$\alpha = \frac{d(r+b)}{2ra}$$

Substituting this expression for α into Equation 5.2 gives:

$$c = \frac{4\pi nr}{\alpha} = \frac{4\pi nr \times 2ra}{d(r+b)} = \frac{8\pi nr^2 a}{d(r+b)} \tag{5.3}$$

In Foucault's first experiments, OC was 4 m but was later increased to 20 m. His value for c was 298,000 km s^{-1}, the chief source of error being in the measurement of the image displacement d which was only 0·7 mm. In 1862, Fizeau determined the velocity of light in water; the path length was 20 m and the mirror speed 800 rev s^{-1}. The experiment was repeated by Cornu and modified by Newcomb, who used a square polished steel prism in place of the mirror M, and by Michelson, who increased the path length OC so that d was 13·3 cm.

Michelson's work on the determination of the velocity of light in air extended over more than 40 years culminating in his experiments of 1924–26, in which a rotating octagonal mirror was used, and in the experiments of 1929–31, in which the light travelled *in vacuo* and which were completed after his death.

5.5 The Determination of the Velocity of Light by Michelson: Use of a Rotating Octagonal Prism

Improvements over a period of 40 years on Foucault's original apparatus led to the now classic experiments of 1924–26 undertaken by Michelson. The light path was increased to 44 miles, while the speed at which the mirror needed to be rotated was effectively reduced by the use of a perfectly made octagonal prism with reflecting sides; the angles between the facets of the prism were accurate to 1×10^{-6} rad. The inconvenient measurement of d, the image displacement, was avoided by adjusting the speed of rotation of the octagonal prism so that light initially reflected from one face was, on return, reflected from the adjacent face; thus, the images were made to coincide.

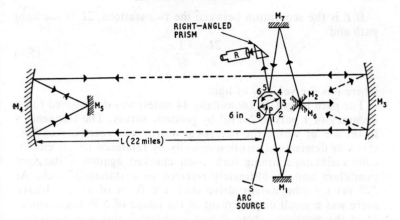

Fig. 5.10. The rotating octagonal prism method of Michelson

In the rotating prism method (Fig. 5.10), the light from the slit S (illuminated by an arc source) travels to one face of the octagonal prism P (of silvered glass or polished stainless steel) and is reflected to the plane mirror M_1; from there, it is reflected to M_2 (M_2 is above the plane of the diagram, and mirror M_6 is below) on to the paraboloidal concave mirror M_3 (focal length, 30 ft; aperture diameter, 2 ft) and emerges from M_3 in an accurately collimated beam, S being at the principal focus of M_3. From M_3, the light travels to the distant mirror M_4 (identical to M_3), where it is brought to a focus at a small concave mirror M_5. The mirror M_5 returns the light back via the incident path to M_3 and then via mirrors M_6, M_7 and the opposite face of the octagonal mirror P to form an image in a low-power telescope R of the slit S.

In Michelson's original experiment, the mirrors M_4 and M_5 were on the summit of Mount San Antonio in California, USA. The observer and the rest of the apparatus were 22 miles away, on the summit of Mount Wilson.

During the initial adjustments, the octagonal mirror P is stationary and the image position in the telescope eyepiece is noted. The octagon is then rotated at a speed of f rev s^{-1} such that the light which was reflected from face 5 (opposite face 1) of the octagonal mirror is now reflected from the adjacent face 4. This occurs when the mirror turns through exactly one-eighth of a revolution during the time that it takes for the light to travel to M_4 (22 miles away) and back. (Light takes approximately 2.3×10^{-4} s to travel 44 miles.) The slit image is then formed in the same position in the telescope eyepiece as if the octagonal mirror were stationary.

If L is the separation between the two stations, $2L$ is the light path and

$$\frac{2L}{c} = \frac{1}{8f} \qquad (5.4)$$

where c is the velocity of light.

The path length $2L$ (nominally 44 miles) was determined to an accuracy of 1 part in 2×10^5 by geodetic survey. The frequency f in rev s^{-1} at which the octagonal mirror was rotated was 528; this was determined stroboscopically by reference to an electrically maintained tuning fork, itself checked against a standard pendulum and so ultimately referred to a standard clock. At 528 rev s^{-1} (the mirror drive took the form of an air blast), there was a small displacement of the image of S in the eyepiece from the position where P was stationary; this was measured against the eyepiece graticule scale and a correction made.

As in all the terrestrial experiments described so far, the velocity of light in the atmosphere is found first. To obtain the velocity c in free space, a correction has to be made for the refractive index of air relative to a vacuum. In Michelson's experiment, this refractive index could not be determined as accurately as the path length because of non-homogeneous atmospheric conditions over the 22 mile path. The velocity of light in free space, c, was given by Michelson to be $(299{,}796 \pm 3)$ km s^{-1} from 200 observations ranging between 299,756 km s^{-1} and 299,803 km s^{-1}.

Dissatisfied with this remarkable work because of inaccurate knowledge of the refractive index of the atmosphere along the light path (which depends on the air pressure, temperature and humidity), Michelson planned the *one-mile evacuated pipe experiment*. This was similar to the rotating octagonal prism method but the light travelled through a pipe 3 ft in diameter, 1 mile in length, and evacuated by mechanical vacuum pumps to a pressure between 0·5 mm Hg and 5·5 mm Hg. The work was completed by Pease and Pearson after Michelson's death in 1931. In this experiment of Michelson, Pease and Pearson, a light path of almost 13 km was achieved. So that the shorter time required for the light to travel this distance—compared with the 70 km (44 miles) of the experiment in air—could be determined, the octagonal mirror was replaced by one with four times the number of facets (i.e. a prism with 32 facets was used).

Light from the vertical slit S (behind which is an arc source) is reflected from one face of the 32-sided mirror P through a window into the evacuated pipe to reach the mirror Q (see Fig. 5.11). From Q, the light travels to a paraboloidal concave mirror N of focal length 50 ft and aperture diameter 40 in. As slit S is at the

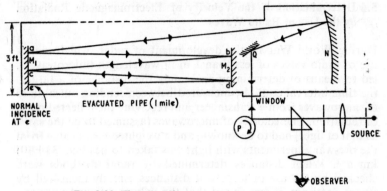

Fig. 5.11. The one-mile evacuated pipe experiment

principal focus of N, light from N is reflected in a parallel beam to the plane mirror M_1, 1 mile away. Reflection then occurs between M_1 and a plane mirror M_2; the mirrors M_1 and M_2 are nominally 1 mile apart and slightly inclined to one another. If M_1 and M_2 are mounted very precisely, reflections take place at a, b, c, d and finally at e where it is arranged that incidence is normal. At e, therefore, the light is returned back along its incident path to the eyepiece of the observer's telescope. The right-angled prism avoids the necessity of the observer and source being in the same line.

The mirror P is rotated by an air-blast drive; the speed of rotation is measured at which an image of S appears in the eyepiece in the same place (or at a small measured displacement from) the image position when P is stationary. The calculation is as before, but Equation 5.4 now becomes

$$\frac{L}{c} = \frac{1}{32f}$$

L being the total length of the light path.

A total of 2,885 determinations were made. After correction for the residual gas in the pipe, c was given as $(299{,}774 \pm 11)$ km s^{-1}. However, this value is lower than recent determinations, and the systematic error has been attributed to instability in the path length in the pipe or to the refraction in the parts of the light path not in the vacuum. A plot of the 2,885 values obtained showed a non-Gaussian distribution, so the quoted probable error of ± 11 km s^{-1} overestimates the accuracy. The evacuated pipe was on unstable alluvial soil near the coast, and a correlation between the fluctuations in the recorded values of c and the state of the tide has been reported.

5.6 Determination of the Velocity of Electromagnetic Radiation by the Use of Radio Waves

During World War II, the development of radar, which makes use of radio waves of less than 1 m in wavelength (microwaves), led to means of determining accurately the distance of a target: the time interval could be measured between sending out a pulse of microwaves from a transmitter and receiving the reflected pulse of radiation. The velocity of microwaves (assumed to be the same as that of light) had to be known, and a weighted mean value from the pre-war experiments with light was taken to be $(299,733 \pm 10)$ km s^{-1}. When distances determined by radar methods were checked by the use of targets at distances directly measured by survey methods, it was found that the values obtained were too small by about 1 part in 20,000. This was not a serious error in the location of aircraft distances, for example, but it was surprising as being considerably greater than the estimated errors attributed to the work of Michelson and other pre-war experiments. Nevertheless, it indicated that the velocity of electromagnetic radiation *in vacuo* was 15 km s^{-1} greater than the values determined for light.

Observations of variable stars did not support the view that the velocity depended upon the wavelength of the electromagnetic radiation. If different colours (wavelengths) of light were propagated with different velocities, the apparent colour would change as the star waxed or waned. Light from many of the variable stars takes several centuries to reach the Earth; even a small difference in velocity would be detectable by apparent colour change on Earth.

Further evidence that velocity is not dependent on wavelength became available in 1951, when an intense source of radio waves was observed (in radio astronomy) to come from a region where two distant galaxies collided, the collision being seen in photographs made with the 200 in telescope at Mount Palomar in California. The Doppler shift in the visible spectrum gave the velocity of the galaxy relative to Earth, and this velocity is proportional to the distance from Earth. The collision was thus found to have taken place $2 \cdot 7 \times 10^9$ light years away from Earth. A similar experiment on the Doppler shift in the radio spectrum at a wavelength of 21 cm for the hydrogen line gave the same distance.

In the laboratory, high-precision radio frequency methods of determining the velocity of electromagnetic radiation are based on the independent determination of the frequency and the wavelength of electromagnetic waves. The velocity c is equal to the frequency multiplied by the wavelength. Amongst several experiments those of Essen, who made use of a cavity resonator, and of Froome, who employed microwave interferometry, are noteworthy.

The *cavity resonator experiments* of Essen were begun in 1949 and made use of an improved version of the Essen and Gordon–Smith apparatus of 1947; basically, the axial length of a cavity resonator *C*, in the form of a cylinder, can be varied by a plunger *P* (Fig. 5.12). Such a cavity resonator with metallic conducting

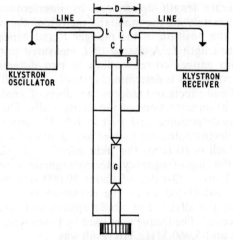

Fig. 5.12. Essen's cavity resonator method for determining the velocity of electromagnetic radiation in vacuo

walls can support standing electromagnetic waves of specific wavelengths in various resonant modes, each of which corresponds to a definite resonant frequency. It can be shown that a particular resonant frequency *f* multiplied by an expression for the wavelength *λ* (which depends on the diameter *D* of the cylindrical cavity, its length *L*, the Q-value of the cavity and a geometrical factor depending on the mode of excitation) will give *c*, the velocity of electromagnetic radiation in free space, provided that the cavity is evacuated to a low pressure. The cavity dimensions are of the order of 10 cm—conveniently small and so leading to easily maintained constant conditions of pressure and temperature.

The particular mode of excitation of standing electromagnetic waves in the cavity is generated by coupling the output of a klystron oscillator by a coaxial line to a loop *l* in the cavity. The fact that resonance is achieved is measured by a maximum signal obtained at a klystron detector and receiver. The frequency f of the klystron oscillator is varied to give this resonance, and f is measured by reference to the known harmonics of the output of a standard oscillator, itself ultimately calibrated against a standard clock. The wavelength λ is calculated from the cavity dimensions, knowledge of the particular mode of excitation, and the Q-value; the latter quantity is both calculated (it compares with $\omega L/R$ of a tuned LCR circuit) and separately measured by experiment. Hence c is found.

The accurate length determined by interferometry (and so referred ultimately to the standard metre) of the slip gauge G determines the position of the plunger P and hence the change of the cavity length L. A change of L measured in terms of two different slip gauges corresponding to two different measured resonant frequencies is determined instead of L itself; this eliminates small end effects and reduces the effects of small mechanical and electrical imperfections of the cavity walls. The diameter of the cavity is determined to 1 part in 10^6. The inner walls of the cavity are electroplated with silver to a depth of 5 μm (1 μm = 10^{-6} m), which is 10 times the 'skin effect' depth (i.e. the depth to which the high-frequency electromagnetic waves penetrate the walls). Thus, a Q-value of about 50,000 is achieved; this is 75% of the calculated value, the reduction being due to slight oxidation of the silver. The whole apparatus is enclosed in an evacuated vessel. The frequencies used by Essen were 9,000 MHz, 9,500 MHz and 5,960 MHz. His result was

$$c = (299{,}792{\cdot}5 \pm 1{\cdot}0)\ \text{km s}^{-1}$$

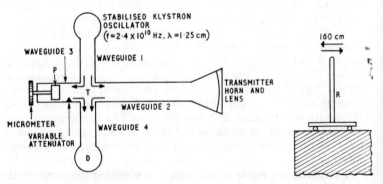

Fig. 5.13. Froome's method of determining the velocity of electromagnetic radiation by the use of a microwave interferometer

The use of *microwave interferometry* to measure the velocity of electromagnetic radiation was undertaken by Froome in 1950 (Fig. 5.13). A stabilised klystron operating at a measured frequency of $2 \cdot 4 \times 10^{10}$ Hz emits radiation of wavelength nominally $1 \cdot 25$ cm. The frequency is accurately known, and the microwave interferometer is used to measure accurately the corresponding wavelength λ of the electromagnetic radiation in air. The velocity of the electromagnetic radiation in air is $f\lambda$ which, when corrected for the refractive index of air, gives c the velocity in free space.

The electromagnetic radiation excited by the klystron oscillator passes through the waveguide 1 to the hybrid junction T, where its energy is equally divided to left and right in the diagram. The left-hand radiation via waveguide 3 is reflected at the plunger P, whilst the right-hand radiation via waveguide 2 is collimated by the transmitter horn and lens to pass through air to the reflector R. This reflector is a silver-plated copper sheet, freely movable to and fro along the prolonged axis of waveguide 2. After reflection of these two waves, they return to the hybrid junction T where they divide again into two parts, one upwards and the other downwards. The downward waves interfere in traversing the waveguide 4 and produce a signal output from the klystron detector and receiver D. When the phase difference between the two waves reaching D is 180°, a minimum signal is obtained from D. The reflector R is moved to achieve this anti-phase condition. Movement of the reflector R through $1 \cdot 6$ m gives 250 successive minima; thus $1 \cdot 6$ m corresponds to a distance of $125 \cdot 5\ \lambda$, where λ is the wavelength in air. The plunger P is adjusted in conjunction with the variable attenuator in waveguide 3 to equalise the amplitudes of the interfering waves which reach D, and so ensure well-defined minima.

After correction to free space, Froome's result was

$$c = (299{,}792 \cdot 6 \pm 0 \cdot 7) \text{ km s}^{-1}$$

A year later, Froome used a considerably modified, improved apparatus from which he obtained the result $c = (299{,}793 \cdot 0 \pm 0 \cdot 3)$ km s^{-1}.

5.7 The Velocity of Light in Material Media

As mentioned in Section 5.4 a main purpose of Foucault's experiments using a rotating mirror was to measure the velocity of light in a medium optically dense compared with air. At that time (1850), this experiment was of importance in deciding between

Newton's corpuscular theory of light and Huygens' wave theory. According to the former, refraction was said to be due to attractive forces between the matter and the light corpuscles; this theory lead to the conclusion that light should travel with greater velocity in the optically denser medium—indeed that the velocity was proportional to the refractive index. The wave theory, on the other hand, established that the velocity of light in matter should be inversely proportional to the refractive index. Thus, in water with a refractive index of 1·33, the Newtonian theory predicts a velocity of 1·33c, whereas the wave theory predicts a value of $c/1·33$.

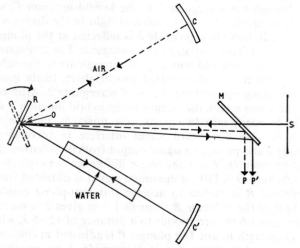

Fig. 5.14. Foucault's apparatus for the determination of the velocity of light in water

The experiments undertaken by Foucault were only qualitative but indicated the correctness of the wave theory. Later experiments by Michelson in 1885 with similar apparatus were quantitative and showed that the velocity was $c/1·33$ in water and $c/1·758$ in carbon disulphide, thus confirming the wave theory.

In Foucault's arrangement (Fig. 5.14), parallel light from the slit S traverses a semi-transparent (semi-silvered) mirror M to the axis of rotation O of the rotating front-surface mirror R. Depending on the position of R, light is reflected from the similar spherical concave mirrors C or C'. Between C and R the light path is in air; between C' and R the light path is largely through water. The light reflected back from C or C' returns to R (where O is at the centres of curvature of both C and C') and back to the

semi-transparent mirror M. During the times taken for the light
to travel the distances OCO and $OC'O$ respectively ($OC = OC'$
geometrically), the mirror R rotates through an angle α; the light
returning to the semi-transparent mirror then reaches P if the
container in the beam RC' is filled with air, whereas this returning
light is deflected to P' if the container is filled with water. The
greater displacement to P' with water showed that the velocity of
light in water was smaller than in air.

Michelson's experiment of 1885 confirmed the relationship

$$n = \frac{\text{velocity of light in free space}}{\text{velocity of light in a transparent material}}$$

where n is the refractive index of the material relative to a vacuum
(taken as unity). Michelson also observed that the use of a white-
light source gave rise to a short spectrum showing the dependence
of velocity on colour (i.e. on wavelength), and thus demonstrating
dispersion.

Several subsequent experiments have been undertaken on the
velocity of light in transparent media. In particular, in 1954,
Bergstrand developed a method due to Houston to measure the
velocity of light in glass and in calcite.

5.8 Group and Phase Velocity

If n is the refractive index of a material relative to that of a
vacuum (for which $n = 1$), the velocity of light in that medium
will be c/n, where c is the velocity of light *in vacuo*. However,
if the light concerned is not monochromatic, various wavelengths
will be present and, since n is a function of wavelength, there
will be dispersion. In a dispersive medium such as air, the velocity
will, therefore, vary with the wavelength of the light used. For
dry air at standard temperature and pressure (s.t.p.: 0°C and
760 mm Hg), the refractive index is 1·000292 for light of wavelength 5,893 Å (the sodium D line).

The refractive index of dry air at s.t.p. for a wavelength λ is
given approximately by the equation

$$n = 1 \cdot 000287 + \frac{1 \cdot 625 \times 10^{-14}}{\lambda^2}$$

where λ is in centimetres. Therefore,

$$\frac{\mathrm{d}n}{\mathrm{d}\lambda} = -\frac{2 \times 1 \cdot 625 \times 10^{-14}}{\lambda^3} = -\frac{3 \cdot 25 \times 10^{-14}}{\lambda^3} \qquad (5.5)$$

In the measurement of the velocity of light in a terrestrial experiment in which the path is through air, the error due to this dispersion is significant.

This error is a result of the difference between the group and the phase (or wave) velocity of the light. Consider Fig. 5.15(a), in which the full line represents a wave of wavelength λ; the dotted line represents a wave of wavelength λ' which traverses the same path, where $\lambda' > \lambda$. It is supposed that at B the waves are in step and reinforce one another (they are not truly in phase as the frequencies are different). At A and C, they are out of step: the displacement due to the wave represented by the full line is in the opposite direction to that represented by the dotted line. The combination of these two waves is therefore a train of waves of amplitude varying from zero at A to a maximum displacement at B and to zero again at C [Fig. 5.15(b)].

If these two waves were propagated with the same velocity, the group or train of waves would move forward preserving the

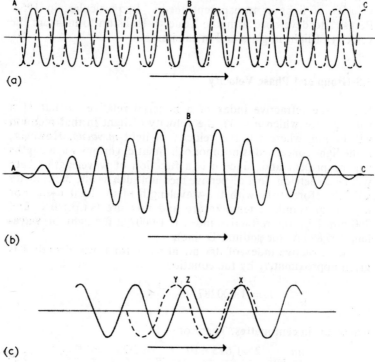

(a)

(b)

(c)

Fig. 5.15. Group and phase (wave) velocity

form shown in Fig. 5.15(b). If light is travelling in a dispersive medium, radiation of longer wavelength will have the smaller refractive index and the larger velocity. Hence, the waves of wavelength λ' will have a velocity v', which exceeds v (the velocity of the waves of wavelength λ). In Fig. 5.15(a), therefore, the wave represented by the dotted line will travel to the right at a greater speed than the wave represented by the full line. The result will be that the waves get more out of step at C as time passes and more in step at A. This is equivalent to the position of maximum amplitude at B moving backwards to the left in Fig. 5.15(b). As this maximum B is the peculiarity in the wave which is observable experimentally, the group of waves will thus have a group velocity v_g less than the velocity v of the individual wave of wavelength λ, v being the phase (or wave) velocity.

In Fig. 5.15(c), let the wave of wavelength λ' move with velocity $v' > v$, the velocity of the shorter wave of wavelength λ. The 'centre' of the group, defined as the point of maximum amplitude, is originally at X. Let T be the time required for the peak marked Y, moving with velocity v', to overtake the peak marked Z. The velocity of approach of Y to Z is clearly $v' - v$. Therefore,

$$(v' - v)T = \text{distance } YZ = \lambda' - \lambda$$

If $v' - v = \Delta v$ and $\lambda' - \lambda = \Delta\lambda$, then

$$T = \frac{1}{\Delta v / \Delta\lambda} \tag{5.6}$$

During this time T, the wave of wavelength λ will move forward a distance vT. The centre of the group, initially at X, then at Z, will have moved in this time T a distance $vT - \lambda$; so the group velocity will be:

$$v_g = \frac{vT - \lambda}{T}$$

Therefore

$$v_g = v - \frac{\lambda}{T}$$

Substituting for T from Equation 5.6,

$$v_g = v - \lambda \frac{\Delta v}{\Delta\lambda}$$

which, in the limit, becomes:

$$v_g = v - \lambda \frac{dv}{d\lambda} \tag{5.7}$$

This gives the relationship between the group velocity v_g and the phase (or wave) velocity v. For a medium free from dispersion, $dv/d\lambda = 0$ and $v_g = v$. In free space, therefore, the group velocity and the phase velocity of light are equal.

In dry air at s.t.p., however,

$$v_g = \frac{c}{n} - \lambda \frac{d}{d\lambda}\left(\frac{c}{n}\right) = \frac{c}{n} + \frac{\lambda c}{n^2}\frac{dn}{d\lambda}$$

If n is put as approximately equal to one, the error due to neglecting the dispersion term is seen to be $100\,\lambda\,(dn/d\lambda)\%$. At a wavelength $\lambda = 6\times10^{-5}$ cm, since $dn/d\lambda = -3\cdot25\times10^{-14}/\lambda^3$ from Equation 5.5, the error is:

$$\frac{600\times10^{-5}\times3\cdot25\times10^{-14}}{6^3\times10^{-15}} = 9\times10^{-4}\%$$

At a nominal velocity of 3×10^5 km s^{-1}, the error due to neglect of dispersion—and so the difference between the group velocity (which is measured in air) and the phase velocity—is, therefore, approximately 3 km s^{-1}.

It is surprising that this source of error in terrestrial measurements of the velocity of light in air, as in the classical experiments of Michelson (Section 5.5) was neglected until it was pointed out by Birge in 1941.

5.9 The Importance of an Accurate Knowledge of c, the Velocity of Light in Free Space

A knowledge of the value of c is manifestly important in terrestrial and astronomical measurements of distance by means of light and radio waves. It is also vital in relating the frequency f and wavelength λ of all forms of electromagnetic radiation by the equation $c = f\lambda$, and in the consideration of the velocity of electromagnetic radiation in media of refractive index n.

The importance of c is furthermore of fundamental importance in many equations in physics, particularly those involving the interaction of radiation with matter and the emission of radiation by matter and those concerned with relativity.

The study of the interaction of electromagnetic radiation in the form of infra-red radiation, visible light, ultra-violet radiation, X-radiation and gamma-radiation with atoms and molecules involves quantum mechanics, in which is implicit the concept of the *photon* with an energy $h\nu = hc/\lambda$ and momentum $h\nu/c$, where

h is Planck's constant and ν is the frequency of the radiation of wavelength λ.

The theory of relativity involves the fundamental postulate of Einstein that the velocity of light in free space is invariable and is the largest possible real velocity in the universe. From the special theory of relativity are deduced the equations

$$l = \frac{l_0}{\sqrt{(1-\beta^2)}}$$

i.e. a rod moving longitudinally through an inertial frame with a velocity v is shortened relative to that frame by a factor

$$\gamma = \frac{l}{l_0} = \frac{1}{\sqrt{(1-\beta^2)}}$$

where $\beta = v/c$. The rate of a clock moving with a speed v is decreased by the same factor γ, and the mass of a body moving at the speed v is increased by the factor γ, or

$$m = \frac{m_0}{\sqrt{(1-\beta^2)}}$$

where m_0 is the rest-mass of the body (i.e. its mass as determined by an observer relative to whom the body is at rest), and m is this mass determined by an observer relative to whom the body is moving at the speed v. Consistent with this last equation is the equivalence of mass m and energy E according to the formula

$$E = mc^2$$

Thus, in the many equations concerned with fundamental processes in science, the velocity of light c is involved, and an accurate knowledge of its value is of prime importance.

Exercise 5

1. Give an account of one precise optical method for measuring the velocity of light. Discuss the significance of this quantity in Physics. (L. G.)

2. Describe a modern optical method of measuring the velocity of light. Discuss any corrections that must be applied to obtain the velocity in vacuo when the experiment is performed in air.
   Are there any reasons to suppose that the velocity of visible light waves should differ from the velocity of electromagnetic radiation measured in a cavity resonator? (L. P.)

3. Give a detailed description of a precision electro-optical shutter method for determining the speed of light.

Discuss the principal corrections which must be applied to obtain the velocity in vacuo when the experiment is performed in air. (L. P.)

4. Discuss the distinction between *phase* and *group* velocities and derive an equation relating them. What bearing has the distinction on the determination of the velocity of light?

Calculate the group velocity of light of wavelength $\lambda = 5 \times 10^{-5}$ cm in carbon disulphide for which

$$\mu = 1 \cdot 577 + \frac{1 \cdot 78 \times 10^{-10}}{\lambda^2}$$

$(c = 3 \times 10^{10}$ cm per sec). (L. P.)

5. Describe a method for determining *either*

(a) the velocity of visible radiation by a method involving the use of an electro-optical shutter, or

(b) the velocity of microwave radiation.

In what respect does the correction to vacuo for the velocity of light differ from that for the velocity of microwaves determined from cavity resonator experiments? (L. P.)

6. Survey the historical development of methods of measuring the velocity of light. (L. P.)

7. Distinguish between wave velocity and group velocity and establish a relationship between them.

The speed of sodium yellow (D lines) in a certain material is to be determined by measuring the time a light pulse takes to travel a known distance. The refractive index of the material for the sodium D lines is:

Wavelength (λ)	Refractive index (n)
5,890 Å	1·63510
5,896 Å	1·63500

Compute the expected answer. (M. P.)

CHAPTER 6

Photometry

6.1 Introduction

Photometry is concerned with the measurement of 'quantity' of light or, more usually, with the time rate of flow of light. Examples are the measurement of the rate of flow of light emitted by a luminous source or an illuminated object, the amount of light received per unit time by a surface, the light reflected (specularly or diffusely) from a surface, and the transmission of light through a material.

A source of light will radiate with a certain distribution of energy in the spectrum. For example, a tungsten filament lamp at, say, 3,000°K, will radiate energy of which a good deal is in the infra-red region of the spectrum. Photometry concerns that aspect of radiant energy to which the human eye responds. Most of the radiated energy from the lamp will be outside the spectral response of the eye. It is clearly essential, therefore, to distinguish between *radiant energy* Q and luminous energy Q_v, the suffix v denoting 'visible'. Often, Q and Q_v will be considerably different since Q_v is only that part of Q to which the human eye responds.

The terms involved in radiation are the radiant energy Q, expressed in ergs or joules, and the *radiant flux* $\Phi = dQ/dt$, expressed in ergs per second or joules per second (i.e. watts). The radiant flux emitted per unit area of a source is the *radiant emittance* $M = d\Phi/dA$ and is expressed in watts per square centimetre. The radiant flux received per unit area is the *irradiance* $E = d\Phi/dA$, also expressed in watts per square centimetre. The *radiant intensity* $I = d\Phi/d\omega$, is expressed in watts per steradian (W sr^{-1}), where ω is the solid angle through which the flux is

167

radiated. (One steradian is the solid angle subtended at the centre of a sphere by an area on the surface of the sphere equal to the square of the sphere's radius.)

The *luminous energy* Q_v, *luminous flux* Φ_v, *luminous intensity* I_v and other terms to be introduced later are all concerned with the visual sensation produced. The response of the human eye to light is at a maximum at 5,550 Å in daylight for the 'normal' eye, and this maximum shifts towards the blue to become about 5,100 Å at low light intensities. The extreme ends of the range at which visual response falls to zero are at 3,800 Å and 7,600 Å. The form of the visual response curve for the eye is well established (see Fig. 6.5) and can be expressed as V, a function of wavelength λ, where the function V is usually considered graphically.

To obtain a suitable unit for luminous flux and other quantities in photometry, a standard source of known distribution of energy in the spectrum is obviously necessary. This is best related to a black-body at a defined temperature because the distribution of radiant energy in the spectrum from a black-body is known fundamentally. The black-body primary source of light is described in Section 6.2.

The human eye, despite its subjective nature as a physical measuring instrument, is nevertheless capable of distinguishing about 180 different hues. Its response to sensation is logarithmic, like that of other human senses. The pupil contracts in bright light to protect the eye, whilst the retina sensitivity increases on secretion of visual purple in poor light. The eye can thus operate comfortably over an enormous range of intensity but this miraculous adaptability prevents the estimation of a numerical value of the light intensity.

In photometric measurement, a great asset of the eye is that it is able to judge accurately when the illumination of two surfaces is identical provided that the illumination is neither too great nor too small; an optimum value of illumination for this purpose is about 30 lm ft^{-2} (see page 170 for the definition of the lumen).

6.2 Standard Sources and Luminous Intensity

The original standard source of light was a candle made from a specified wax and wick. This is unsuitable for precise specification because the light emitted from identical candles differs by as much as 20% of their mean. It was therefore superseded by the Vernon–Harcourt pentane lamp but this in turn became inadequate with the increasing demand for accuracy—largely as a result of the introduction of electric lighting.

The modern standard source is a black-body in the form of a small hollow cylinder of pure fused thoria, of internal diameter 2·5 mm and length 45 mm, immersed in platinum maintained at its freezing point (1,773°C = 2,046°K). The main features of this source are shown in Fig. 6.1. A surface area of 1 cm² of platinum at 1,773°C arranged as a black-body (full) radiator has a candle power (based on the previous pentane lamp standard) of 58·9. The present internationally accepted standard of *luminous intensity* is the *candela* (the Italian word for candle), taken to be the luminous intensity of a source which emits light energy at 1/60 (60 is the nearest convenient integer to 58·9) times the rate at which luminous energy is emitted by an area of 1 cm² of a full radiator at the freezing point of platinum.

Luminous intensity I_v (or candle power) is thus specified in candelas (abbreviation, cd).

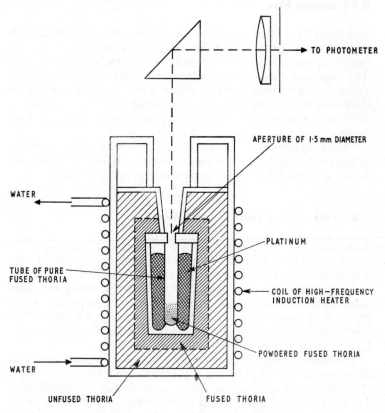

Fig. 6.1. The black-body primary source of light

In practice, tungsten filament lamps are used as sub-standards. These have the advantage that they can be run at a high operating temperature of about 3,000°K because the melting point of tungsten is 3,653°K. One temperature standard for calibration is 2,854°K. This permits a high light emission without undue deterioration of the filament because of evaporation of the tungsten in the evacuated lamp envelope.

The sub-standard filament lamps are calibrated against the primary molten platinum full radiator by national standards laboratories (National Physical Laboratory in Britain). The standard of accuracy required in photometry necessitates the control of the potential difference across the sub-standard lamp filament to 0·1 V at 200 V.

6.3 Luminous Flux

Luminous flux is that quantity characteristic of radiant flux which expresses its capacity to produce visual sensation. The luminous flux Φ_v is the rate of flow of luminous energy, which is dQ_v/dt. The luminous flux emitted within unit solid angle (1 sr) by a point source of one candela which emits light equally in all directions (i.e. is omnidirectional) is called the *lumen* (abbreviation, lm). Manifestly, an omnidirectional small source of 1 cd emits a total luminous flux of 4π lm.

Luminous intensity I_v in candelas can alternatively be expressed in lumens per steradian. For non-uniform radiation, $I_v = d\Phi_v/d\omega$, where ω is the solid angle through which flux is radiated.

6.4 Luminous Flux Density and Luminance

Luminous flux density at a surface is concerned with either the luminous flux emitted per unit area by a luminous source, or with the luminous flux received by unit area of a surface.

In the first case, the specific term is the *luminous emittance* (also known as the *emitted luminous excitance*) M_v, where $M_v = d\Phi_v/dA$, A being area; it is quoted in lumens per square foot (lm ft^{-2}), lumens per square centimetre (lm cm^{-2}) or lumens per square metre (lm m^{-2}).

The second case is concerned with *illumination* (also known as *illuminance*) $E_v = d\Phi_v/dA$ and is quoted in lumens per square foot. The lumen per square foot is the same as the *foot-candle* (fc), i.e. the illumination produced at a surface by a source of 1 cd at a distance of 1 ft. Alternative units of illumination are the

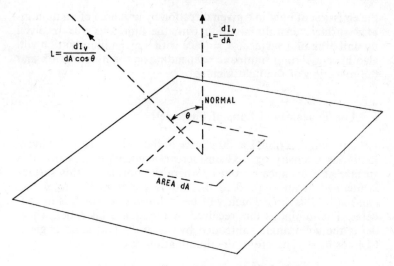

$$L = \frac{dI_v}{dA \cos \theta}$$

$$L = \frac{dI_v}{dA}$$

NORMAL

θ

AREA dA

Fig. 6.2. Luminance and direction of reflected light

lumen per square centimetre (known as the *phot*), and the lumen per square metre (known as the *lux;* abbreviation, lx).

To take into account that a surface will reflect and scatter light incident upon it, and that this reflected light will be in all directions away from the surface unless the surface is optically polished (when specular reflection is concerned), the term *luminance L* (also known as *photometric brightness*) is used. Luminance is defined for a given point on the surface and for a given direction away from the surface as the luminous intensity per unit area of the surface projected perpendicular to the direction concerned. Thus, a small area dA around a point on the surface which produces a luminous intensity of dI_v in a direction normal to the surface will have a luminance dI_v/dA, whereas in a direction at an angle θ to the normal the luminance will be $dI_v/(dA \cos \theta)$ (Fig. 6.2).

The units of luminance are consequently the candela per square centimetre (cd cm^{-2}) also known as the *stilb;* the candela per square metre (cd m^{-2}) or *nit*, and the candela per square foot (cd ft^{-2}). For example, for a tungsten filament at 2,700°K, the luminance is 10^7 cd m^{-2}; for white paper in sunlight, 25,000 cd m^{-2}; and in moonlight, 0·03 cd m^{-2}. Luminance is also expressed in units which are $1/\pi$ times the above. These are the *lambert* = $(1/\pi)$ cd cm^{-2}, the *metre-lambert* = $(1/\pi)$ cd m^{-2} and the *foot-lambert* = $(1/\pi)$ cd ft^{-2}.

Note that luminous emittance is concerned with the luminous flux emitted per unit area by a luminous source, luminance with the emission of light in a given direction by unit area of an illuminated surface, and illumination with the luminous flux received by unit area of a surface. A surface with a given illumination will also have a certain luminance depending on the illumination and the reflection of the light incident.

6.5 Two Fundamental Laws of Photometry

Let an omnidirectional point source of light at P (Fig. 6.3) have a luminous intensity of I_v. Consider an element of area dA of a surface at a distance r from P, where the normal to this surface is inclined at an angle θ to the radius vector r. The total flux emitted by P is $4\pi I_v$, which will be in lumens, where I_v is in candelas. The luminous flux received by the area dA is $I_v\, d\omega$, where $d\omega$ is the solid angle subtended by dA at P. This solid angle is $(dA \cos \theta)/r^2$. Therefore, the illumination at dA is:

$$E = \frac{I_v\, d\omega}{dA} = \frac{I_v\, dA \cos \theta}{dA\, r^2} = \frac{I_v \cos \theta}{r^2} \qquad (6.1)$$

This result of the definitions in the preceding sections is in accordance with the two fundamental laws of photometry:

1. The illumination of a surface element by a point luminous source is inversely proportional to the square of the distance between the source and the surface element.

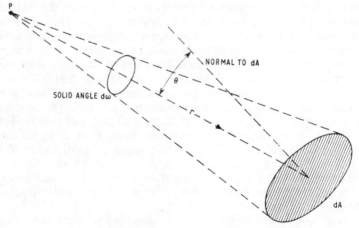

Fig. 6.3. Illumination of a surface element by a point luminous source

2. The illumination is proportional to the cosine of the angle between the normal to the surface and the radius vector joining the source and the surface element.

Equation 6.1 is an expression of *Lambert's cosine law*, which may be said to embody the two fundamental laws of photometry and is consistent with the fact that light is propagated in straight lines in a medium of homogeneous refractive index and free of obstacles.

6.6 Radiant Flux and Luminous Flux

When a lamp or other source of radiant energy Q is used, the eye can only perceive that radiation of wavelength within the visible spectrum (from 3,800 Å to 7,600 Å); i.e. the eye only perceives the luminous energy Q_v. Often the source emits radiation at wavelengths outside this range and only a fraction of the total radiant flux is luminous. Moreover, the stimulus that the eye receives depends on its own variation in response with wavelength.

For a black-body radiator, the distribution of energy in the spectrum is determined by the Planck equation

$$Q_\lambda = \frac{8\pi hc}{\lambda^5[\exp(hc/k\lambda T)-1]} \tag{6.2}$$

where $Q_\lambda \, d\lambda$ is the energy per unit volume of the radiation of wavelength lying between λ and $\lambda+d\lambda$, h is Planck's constant, c is the velocity of light in free space, k is Boltzmann's constant ($1\cdot38\times10^{-16}$ erg deg^{-1}K), and T is the absolute temperature (in degrees Kelvin) of the black-body radiator. Typical plots of Q_λ against λ at various temperatures T are given in Fig. 6.4.

It is seen from Fig. 6.4 that the wavelength λ_m at which the maximum in the distribution curve occurs decreases as the temperature T increases. This maximum is found by differentiation of Equation 6.2 with respect to λ and equating to zero. This gives

$$\lambda_m T = \frac{ch}{4\cdot965\,k} \tag{6.3}$$

a result known as *Wien's displacement law*. Employing c.g.s. units,

$$\frac{ch}{4\cdot965\,k} = \frac{3\times10^{10}\times6\cdot6\times10^{-27}}{4\cdot965\times1\cdot38\times10^{-16}} = 0\cdot29 \text{ cm degK}$$

Therefore,

$$\lambda_m T = 0\cdot29 \text{ cm degK} \tag{6.4}$$

Hence, at $T = 3{,}000°K$, $\lambda_m = 0{\cdot}29/3{,}000 = 0{\cdot}97 \times 10^{-4}$ cm = 9,700 Å, which is in the infra-red.

It follows that a black-body at 3,000°K produces radiation which does not give visible light at the peak of its distribution; only the part of the radiation further towards the shorter wavelengths below 7,600 Å and above 3,800 Å results in luminous flux perceivable by the eye (see the indicated region in Fig. 6.4).

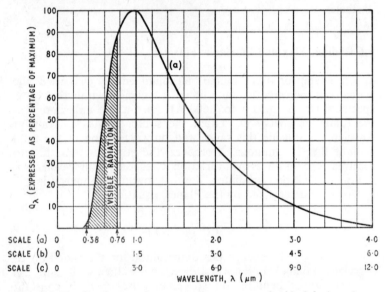

Fig. 6.4. *The distribution of energy in the spectrum of a black-body radiator at various temperatures: (a) $T = 3{,}000°K$, total power radiated = $4{\cdot}62 \times 10^5\,mW\ cm^{-2}$; (b) $T = 2{,}000°K$; (c) $T = 1{,}000°K$. Maximum and total energy $\propto T^4$*

Most light sources are not black-bodies. The familiar tungsten filament in an electric lamp has an emissivity of about 0·4, which varies with wavelength and temperature. (The *spectral emissivity* ε_λ is defined as the ratio of the emission of the object at a wavelength λ to that of a black-body at the same temperature and wavelength.) The tungsten lamp gives a continous spectrum of radiation but much of the electrical power (in watts) fed to the lamp is used in producing heat radiation and not visible light. Energy losses also occur in heating the leads to the filament and the glass envelope. So the luminous flux (in lumens) does not correspond to a fixed number of watts of electrical power supplied to the lamp. For example, a certain tungsten filament lamp emits 12·9 lm W^{-1} at 2,740°K, 15·2 lm W^{-1} at 2,810°K, 18·1 lm W^{-1} at 2,920°K and 21·2 lm W^{-1} at 3,000°K. The practical maximum is about 93 lm W^{-1} for a body at a temperature such as that of the Sun (i.e. 6,500°K). At still higher temperatures, the wavelength at which maximum radiation occurs shifts towards the violet end of the spectrum, and much of the radiation is in the ultra-violet.

It is interesting to observe that the Sun, with a surface temperature of 6,500°K, has a maximum in its spectral distribution curve at a wavelength λ_m given by Equation 6.4 as 0·29/6,500 = 4·5 × 10^{-5} cm, (i.e. 4,500 Å); this is not far from the region at which the sensitivity of the human eye is a maximum. Possibly explained by Darwinian evolutionists!

Sources of light such as discharge lamps containing mercury vapour or sodium will emit line spectra, or, for high-pressure lamps, a line spectrum superimposed on a continuous spectrum. The familiar fluorescent lamps emit a spectrum which is more or less continuous with emphasis in some wave bands, but with a spectral distribution unlike that of a black-body or filament lamp.

6.7 Colour Sensitivity of the Human Eye

The human eye is most sensitive to wavelengths of about 5,550 Å (in the green-yellow), so a small radiant flux at this wavelength will appear as bright as a large flux at, say, 4,400 Å or 6,600 Å.

Suppose a luminous element of area dA emits a radiant flux Φ_λ at wavelength λ. An area of this kind which emitted monochromatic light of wavelength 5,550 Å would produce a radiant flux of $\Phi_{5,550}$ and give a maximum response at the human eye. A second element, of the same area dA, emitting monochromatic radiation of wavelength λ would need to produce a radiant flux of Φ_λ in excess of $\Phi_{5,550}$ to give the same appearance of luminance as the first area. Then $1/\Phi_\lambda = V_\lambda$ is a measure of the sensitivity of the eye to radiation of wavelength λ in producing a given visual response. V_λ is called the *relative luminous efficiency*.

A relative luminous efficiency curve at wavelengths in the visible spectrum for a standard observer has been determined. Difficulties arise in determining this curve, which indicates the power of radiant flux to produce sensations of equal brightness, because at low intensities of illumination the wavelength at which the eye has maximum sensitivity decreases towards 5,100 Å approx.; also, the visual effect depends on the size of the field viewed, to some extent on the duration of the stimulus, and on the part of the retina stimulated. For all practical purposes, the relative luminous efficiency data for normal illumination are used. This illumination is defined as exceeding 100 luxon in a field of view of 2°–3°. The *luxon* is the retinal illumination produced by a surface having a luminance of 1 cd m^{-2} when the pupil area is 1 mm^2.

Values of the relative luminous efficiency V_λ (also known as the *lamprosity*) for wavelengths in the visible spectrum are plotted

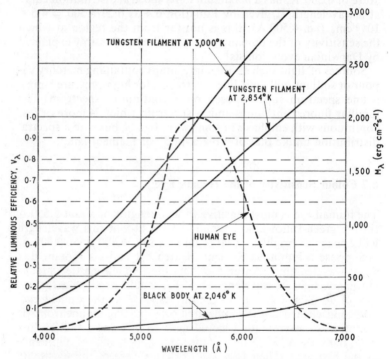

Fig. 6.5. The relative luminous efficiency curves (human eye) and the distribution of energy in the spectrum per unit area of a black-body at 2,046°K and a tungsten filament at 2,854°K and 3,000°K

in Fig. 6.5 relative to a value of unity at the wavelength 5,550 Å; at this wavelength, the human eye has maximum sensitivity when the intensity is high (i.e. under conditions of *photopic* vision), as opposed to low intensity (when the vision is said to be *scotopic*). In photopic vision, the cones in the retina are the sensitive receptors; scotopic vision is by the rods, and the wavelength at which the sensitivity is a maximum shifts to about 5,100 Å.

6.8 Illuminating Power and the Lumen

A fraction of the total radiation from a black-body radiator at the freezing point of platinum (2,046°K) is in the visible region of the spectrum; and, by definition, 1 cm² of this radiator has a luminous intensity of 60 cd. The total radiation at this temperature can be determined (from the temperature rise in a black-body receiver of known thermal capacity subtending at the source a known solid angle), or by calculation using Stefan's law:

$$M = \sigma(T^4 - T_0^4) \tag{6.5}$$

where M is the *radiant emittance* (also known as the *emitted radiant excitance*) integrated over all wavelengths and given in watts per square centimetre, T is the absolute temperature in degrees Kelvin of the black-body, T_0 is the absolute temperature of the surroundings, and σ is Stefan's constant $= 5 \cdot 67 \times 10^{-12}$ W cm⁻² deg⁻⁴ K.

The *luminous efficacy* K of a source is defined as the ratio of the luminous flux to the radiant flux that it emits, which is Φ_v / Φ, and is usually specified in lumens per watt.

Let the luminous efficacy of the black-body source at 2,046°K (the freezing point of platinum) be K_m at the wavelength 5,550 Å (at which the relative luminous efficiency of the radiation is a maximum and equal to unity). The luminous efficacy of this source at any other wavelength will be $K = K_m V_\lambda$, where V_λ is less than unity. The black-body source at 2,046°K produces a radiant flux per square centimetre (radiant emittance) equal to M_λ W cm⁻² in the wavelength interval λ to $\lambda + d\lambda$, M_λ being given by Equation 6.2 as

$$M_\lambda \, d\lambda = \frac{c_1 \lambda^{-5} \, d\lambda}{\exp (c_2/\lambda T) - 1}$$

where $c_1 = 8\pi hc$, $c_2 = hc/k$, and $T = 2,046°K$. The luminous emittance M_v is the product of the radiant emittance and the luminous efficacy. Therefore,

$$M_v = \int_0^\infty K M_\lambda \, d\lambda = K_m \int_0^\infty \frac{V_\lambda c_1 \lambda^{-5} \, d\lambda}{\exp (c_2/\lambda T) - 1}$$

V_λ and M_λ are not simple functions of λ, so graphical or numerical methods of integration are used. The value of this integral in c.g.s. units is $0 \cdot 275$. Hence,

$$M_v = 0 \cdot 275 \, K_m \text{ lm cm}^{-2}$$

By definition, the luminance L of this source is 60 cd cm^{-2}, so the luminous flux $= \pi L$ (see Section 6.12). Therefore,

$$0 \cdot 275 \, K_m = 60 \pi \, \text{lm cm}^{-2}$$

so

$$K_m = 682 \, \text{lm W}^{-1}$$

K_m has been termed 'the mechanical equivalent of light'.

A source of light can give a continuous spectrum which is not the same as that of a black-body (for example, a tungsten filament lamp); if the radiant emittance M_λ is known (usually obtained graphically) as a function of λ, the luminous emittance M_v can be evaluated from the equation:

$$M_v = 682 \int_0^\infty V_\lambda \, M_\lambda \, d\lambda \, \text{lm cm}^{-2}$$

Both M_λ and V_λ are known graphically as functions of wavelength, so graphical or numerical evaluation of the integral can be carried out.

For a source of light, such as a discharge lamp, which emits radiation at discrete wavelengths λ_1, λ_2, λ_3, ..., λ_p at which the corresponding radiant fluxes per unit area (radiant emittances) are M_1, M_2, M_3, ..., M_p W cm^{-2}, the relative luminous efficiencies at these wavelengths are known to be V_1, V_2, V_3, ..., V_p. The total luminous flux per unit area is given by the equation:

$$M_v = 682[V_1 M_1 + V_2 M_2 + V_3 M_3 + \ldots + V_p M_p] \, \text{lm cm}^{-2}$$

The inclusion on Fig. 6.5 of spectral distribution curves for a black-body at 2,046°K and for a tungsten filament at 2,854°K and at 3,000°K shows the marked difference between the manner in which energy is radiated in the visible spectrum by a heated body and the way in which the human eye responds.

6.9 Polar Diagrams

Sources of light do not necessarily radiate equally in all directions. To give one example, a frosted tungsten lamp bulb suspended from a ceiling may radiate fairly uniformly downwards but the light in an upwards direction is restricted by the base of the lamp and its support.

The distribution of light output from a source can be represented by a polar diagram. The luminous intensity I_v in any direction is $d\Phi_v/d\omega$, where ω is the solid angle through which the luminous

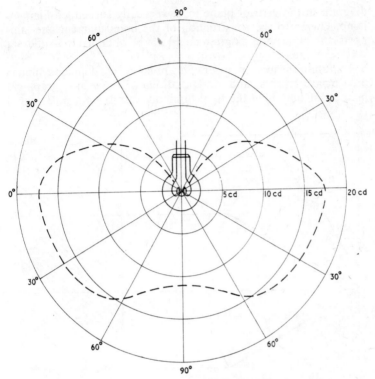

Fig. 6.6. Polar diagram in a vertical plane of a vertically suspended filament lamp

flux Φ_v is radiated. I_v may vary with direction. If a radius vector from the source is drawn in the required direction and of length proportional to the luminous intensity in candelas in that direction, a series of such radii drawn all around the source in a given plane will terminate on the polar curve or diagram of the source. To give a full description of the light distribution, a series of polar diagrams is needed in various planes including the source.

To obtain the polar diagram by experiment, a photoelectric cell is used; the photocell must be far enough from the source to be regarded as a point receiver and arranged with its surface normal to the radius vector from the source.

The *mean spherical candle power* (m.s.c.p.) is the mean of the candle powers, measured in candelas, in all directions about the source as origin. It is the average value of the luminous intensity in all directions. The distribution of light from a source rarely has an exact mathematical form. Fig. 6.6 shows a typical polar diagram in the vertical plane for a vertically suspended filament lamp. Owing to the construction of the lamp filament and supports, the same polar diagram might not be obtained in a different vertical plane.

To obtain the m.s.c.p. by drawing polar curves demands numerous measurements and computations. Rousseau's graphical method of obtaining the total luminous flux from a polar curve

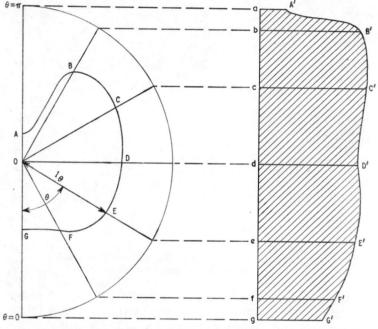

Fig. 6.7. *Rousseau's graphical method of determining the m.s.c.p. of a lamp*

is to draw a circle, which has the source as centre, in the plane concerned (Fig. 6.7). An angular scale is marked on this circle. The angular divisions are projected across to a line drawn parallel to the axis of symmetry of the polar curve. Distances OA, OB, OC, ... are set off as aA', bB', cC', ... as shown. The area shaded in Fig. 6.7 represents the total luminous flux emitted by the source, according to a suitable scale. The m.s.c.p. is then this area divided by the length of the base $ag = 2r$.

To show that this is so, let I_θ be the luminous intensity in candelas in a direction at an angle θ to the axis of symmetry of the polar diagram (Fig. 6.7). The total luminous flux, in lumens, is given by

$$\Phi_v = \int\limits_{\alpha=0}^{\alpha=2\pi} \int\limits_{\theta=0}^{\theta=\pi} I_\theta \sin\theta \; d\theta \; d\alpha$$

where θ varies in the plane of the diagram of Fig. 6.7 from 0 to π, and this plane is considered to be rotated about the axis AOG through an angle α subject to small increments of $d\alpha$; a complete rotation corresponds to the limits of integration being from $\alpha = 0$ to $\alpha = 2\pi$. This expression then gives the luminous flux Φ_v radiated in all directions around the source at O.

Often, the polar diagram in the vertical plane is much the same irrespective of the position of this plane because I_θ does not vary much with α. Then

$$\Phi_v = 2\pi \int\limits_{\theta=0}^{\theta=\pi} I_\theta \sin\theta \; d\theta$$

The Rousseau diagram is drawn so that $OE = eE'$ and similarly for $OA, OB, OC, \dots$ Hence, eE' is drawn to scale to represent I_θ. To find the area of this Rousseau diagram, $aA'G'g$:

$$de = r\cos\theta$$

Therefore,

$$ge = r(1-\cos\theta)$$

so

$$\frac{d(ge)}{d\theta} = r\sin\theta$$

Hence,

$$\text{area } aA'G'g = \int I_\theta \, d(ge) = \int\limits_0^\pi I_\theta r \sin\theta \; d\theta = r \int\limits_0^\pi I_\theta \sin\theta \; d\theta$$

Therefore, the area of the Rousseau diagram multiplied by $2\pi/r$ represents the total luminous flux Φ_v from the source. If the m.s.c.p. of the source is I cd, it follows that:

$$\text{m.s.c.p.} = \frac{2\pi}{r} \times \frac{\text{area of Rousseau diagram}}{4\pi}$$

i.e.

$$\text{m.s.c.p.} = \frac{\text{area of Rousseau diagram}}{2r}$$

6.10 Measurement of Luminous Intensity

The luminous intensities of two sources are conveniently compared by setting them up at opposite ends of an optical bench (of length great compared with the source dimensions) and placing between them a device—*the photometer head*—which they illuminate. The distances of the sources from the head are then varied until the illuminations they produce at the photometer head are equal. Provided that the surfaces of the photometer head are identical and receive light from sources which approximate to points, and that the axis through the centre of the head and the sources is normal to the surfaces of the head

$$\frac{I_1}{r_1^2} = \frac{I_2}{r_2^2}$$

where I_1 is the luminous intensity of source S_1 at a distance r_1, and I_2 is the luminous intensity of source S_2 at a distance r_2. These distances must be measured from the identical screens of the head. If I_1, say, is a standard of known luminous intensity, the candle power I_2 is found.

An accurate form of photometer head for visual observation of equality of brightness of the identical screens is the *Lummer–Brodhun* head. It can only be used if the sources emit light of the same or similar colour; it is unsatisfactory for different colours. The housing of the Lummer–Brodhun head [Fig. 6.8(a)] contains opposite apertures through which light from the sources is admitted. To ensure as a preliminary that the sources S_1 and S_2 are set up accurately collinear with the screen Z, this pair of apertures is frequently provided with a detachable lens and screen assembly.

Light rays from the sources S_1 and S_2, which are to be compared, illuminate opposite sides of a white diffusing screen Z made of magnesium oxide. The diffusely reflected rays from Z travel to the right-angled prisms M_1 and M_2 (mirrors are used as alternatives in some heads), where they are reflected into the Lummer–Brodhun cube consisting of two right-angled prisms P_1 and P_2 contacted at their hypotenuse faces. The prism P_1 has a sandblasted ring on its hypotenuse face. Inside this ring, the faces of P_1 and P_2 are in optical contact but not at the ring itself. Light from M_2 passes straight through the common face of the P_1P_2 combination to be absorbed at the blackened inside walls of the housing *except* where it meets the region of the surface of P_1 which is sand-blasted. At this ring area, the light from M_2 is totally internally reflected from the hypotenuse face of P_2 and

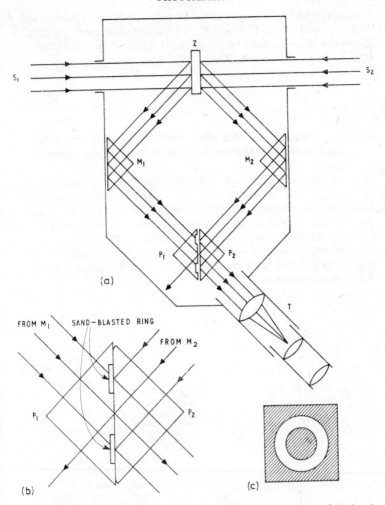

Fig. 6.8. The Lummer–Brodhun photometer head: (a) the housing of the head;
(b) enlarged view of P_1 and P_2; (c) view of P_2 through telescope

enters the viewing telescope T. Light from M_1 which reaches the
back of the sand-blasted ring is scattered and absorbed but light
passing through this ring travels into the telescope T. P_2 is seen
in the telescope T with the illuminations unequal and with the
ring area clearly distinguished [Fig. 6.8(c)]. The distances r_1 and
r_2 of S_1 and S_2 respectively from Z are adjusted to provide
uniform illumination of the entire field so that the ring cannot

be distinguished. The illuminations of the opposite identical sides of the screen Z are then equal. To ensure that these surfaces of Z are the same, Z may be reversed. If I_1 and I_2 are the luminous intensities of the sources, then:

$$\frac{I_1}{I_2} = \frac{r_1^2}{r_2^2}$$

6.11 The Illumination of a Matt Surface Element by a Second Separated Matt Surface Element of Given Luminance

It is an observational fact that self-luminous and reflecting surfaces which are uniformly diffusing have a luminance or brightness which is independent, or almost independent, of the inclination of the surface to the direction of observation. This is nearly true for matt white surfaces (for example, those composed of magnesium oxide and plaster of Paris). A perfect matt surface is one for which the luminous intensity I_v in a direction at an angle θ to the normal is given by

$$I_v = I_0 \cos \theta$$

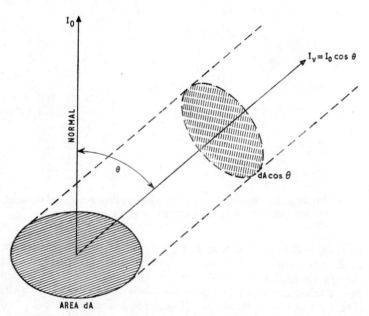

Fig. 6.9. The luminance of a perfectly matt surface is independent of the direction of viewing

where I_0 is the luminous intensity normal to the surface, i.e. it obeys Lambert's cosine law of emission. From Fig. 6.9, it can be seen that the effective area in the direction θ to the normal of an actual diffusing surface of area dA is $dA \cos \theta$. Therefore, the luminance L, or photometric brightness, in candelas per unit area in the direction θ is $(I_0 \cos \theta)/(dA \cos \theta) = I_0/dA$. Thus it is evident that the luminance is the same irrespective of the angle θ.

Consider the illumination provided at a small matt surface of area dS_2 by a second small matt surface of area dS_1 separated from it by a distance r, where dS_1 has a luminance L (Fig. 6.10).

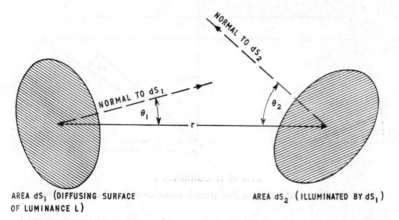

AREA dS_1 (DIFFUSING SURFACE OF LUMINANCE L)　　　　AREA dS_2 (ILLUMINATED BY dS_1)

Fig. 6.10. Illumination at a matt surface element due to luminous flux from a second matt surface element

Let θ_1 and θ_2 be the angles that the normals to dS_1 and dS_2 respectively make with the line joining dS_1 and dS_2. The luminous intensity of dS_1 in the direction of dS_2 is $L \, dS_1 \cos \theta_1$ cd. The solid angle subtended by dS_2 at dS_1 is $(dS_2 \cos \theta_2)/r^2$. Therefore, the luminous flux received by dS_2 from dS_1 is given by the equation

$$d\Phi_v = \frac{L \, dS_1 \, dS_2 \cos \theta_1 \cos \theta_2}{r^2} \quad \text{lm} \quad (6.6)$$

and the illumination dE of dS_2 is given by:

$$dE = \frac{d\Phi_v}{dS_2} = \frac{L \, dS_1 \cos \theta_1 \cos \theta_2}{r^2} \quad \text{lm cm}^{-2} \quad (6.7)$$

6.12 The Total Luminous Flux from a Small Matt Plane Area of Uniform Luminance

If a plane matt white element of area dS has a uniform luminance of L cd cm^{-2}, it will radiate directly into the hemisphere bounded by a spherical surface with dS at its centre and the diametrical plane containing dS (Fig. 6.11). The luminous intensity in the

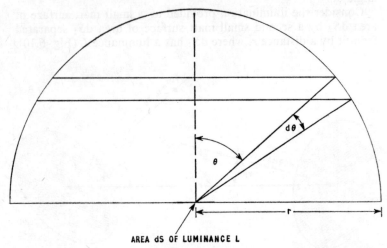

AREA dS OF LUMINANCE L

Fig. 6.11. The total luminous flux from a plane area dS of uniform luminance

direction at an angle θ to the normal to dS is $L\,dS\cos\theta$. This intensity will be the same over a ring-shaped element of area $(2\pi r \sin\theta)r\,d\theta$, where r is the radius of the hemisphere. The solid angle subtended by this ring at dS is $2\pi \sin\theta\,d\theta$, and the luminous flux $d\Phi_v$ radiated into this solid angle is $(L\,dS\cos\theta)$ $2\pi \sin\theta\,d\theta$. Hence, the total luminous flux Φ_v radiated from dS is given by:

$$\Phi_v = \int_0^{\pi/2} 2\pi L\,dS \cos\theta \sin\theta\,d\theta$$

$$= -2\pi L\,dS \int_{\vartheta=0}^{\theta=\pi/2} \cos\theta\,d(\cos\theta)$$

$$= -2\pi L\,dS\,\tfrac{1}{2}\left[\cos^2(\pi/2) - \cos^2 0\right]$$

Therefore,

$$\Phi_v = \pi L\,dS \tag{6.8}$$

6.13 The Integrating Sphere

Instead of finding the m.s.c.p. of a lamp by a series of measurements involving the determination of its polar diagram, an integrating sphere may be used to give this m.s.c.p. from a single measurement. The provision of a small aperture in the sphere's surface and of a lamp of known m.s.c.p. within the sphere provides a useful source of luminous intensity in visual and photoelectric photomery.

A lamp is enclosed in a large sphere of which the inner surface is matt white and which is provided with a small window anywhere in the wall. This window is shielded from direct rays from the lamp by a small matt white screen mounted inside the sphere. The diameter of the sphere may be 6 ft, whereas the dimensions of the lamp and window are not more than a few inches. Light emitted from the lamp in all directions is diffusely reflected or scattered from the walls, and it is clear that rays undergo multiple reflections before emerging from the window. The integrating sphere is illustrated in Fig. 6.12.

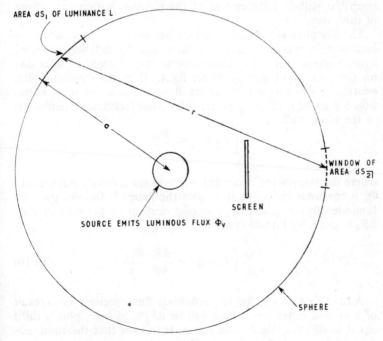

AREA dS_1 OF LUMINANCE L

WINDOW OF AREA dS_2

SCREEN

SOURCE EMITS LUMINOUS FLUX Φ_V

SPHERE

Fig. 6.12. The integrating sphere

Let dS_1 and dS_2 be any small areas of the illuminated inner wall of the sphere. The luminous flux received by dS_2 from dS_1 is given by Equation 6.6, but now the normals to dS_1 and dS_2 are radii of the sphere; therefore $\theta_1 = \theta_2 = \theta$ and $\cos \theta = r/2a$, where r is the separation between dS_1 and dS_2 and a is the radius of the integrating sphere. Therefore, the luminous flux at dS_2 is:

$$d\Phi_v = \frac{L\,dS_1\,dS_2}{4a^2} \tag{6.9}$$

It follows that the luminous flux received from dS_1 by dS_2 is independent of the positions of dS_1 and dS_2 on the sphere's inner surface. This result is not invalidated if light passes from dS_1 to dS_2 via a third surface element dS_3, or even further elements, (i.e. if multiple reflections are involved) because then dS_1 would simply be replaceable by dS_3 or further elements.

It is clear that each part of the inner surface of the sphere will, therefore, be equally illuminated and, as the surface is uniform, each element of it will receive the same quantity of light, which will depend on the total luminous flux Φ_v radiated by the internal lamp but will be independent of the position within the sphere of this lamp.

The illumination at any point on the sphere's inner surface is caused by direct light from the lamp and by diffusely reflected light. Suppose L_D is the luminance of any element of area dS_1 on the inner wall due to direct light. Then the luminous flux emitted by dS_1 owing to its direct illumination is given by Equation 6.8 as $\pi L_D\,dS_1$. If ϱ is the reflectance (reflection coefficient) of the inner wall,

$$\int L_D\,dS_1 = \frac{\Phi_v \varrho}{\pi}$$

where the integration is carried out over the sphere's surface and Φ_v is the total luminous flux from the lamp. It follows that the luminous flux received by a second element of the wall of area dS_2 is given by Equation 6.9 as:

$$\frac{dS_2}{4a^2}\int L_D\,dS_1 = \frac{dS_2}{4a^2}\frac{\Phi_v \varrho}{\pi} \tag{6.10}$$

Added to this will be the luminous flux received as a result of a second reflection, which will be $dS_2\,\Phi_v \varrho^2/4\pi a^2$, plus a third equal to $dS_2\,\Phi_v \varrho^3/4\pi a^2$, and so on. It follows that the luminous flux received by dS_2 owing to scattered light from the rest of the sphere's inner surface is Φ given by:

$$\Phi_{\mathrm{d}S_2} = \frac{\mathrm{d}S_2\,\Phi_v}{4\pi a^2}\,(\varrho + \varrho^2 + \varrho^3 + \ldots) = \frac{\mathrm{d}S_2\,\Phi_v}{4\pi a^2}\left(\frac{\varrho}{1-\varrho}\right) \quad (6.11)$$

In addition to this, there would be the direct luminous flux from the lamp. If, however, $\mathrm{d}S_2$ is the area of an aperture (window) in the wall of the sphere and this window is shielded from direct light from the lamp, Equation 6.11 will give the luminous flux emitted from the window of area $\mathrm{d}S_2$ where ϱ is the reflectance of the sphere's inner wall and Φ_v is the total luminous flux from the lamp. The position of the lamp in the sphere is not important. As $\mathrm{d}S_2$, a and ϱ are all known in practice, measurement of the luminous flux emitted from the window enables Φ_v to be found.

The integrating sphere (a cube is often used as a close practical approximation) is calibrated as a photometer by placing inside it a sub-standard lamp of known total luminous flux Φ_v lm (found by calibration against the standard platinum source by means of a Lummer–Brodhun photometer). The luminous flux $\Phi_{\mathrm{d}S_2}$ through the window is known from Equation 6.11. This is incident upon

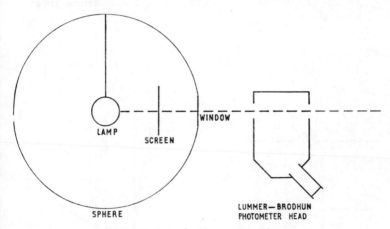

LAMP SCREEN WINDOW

SPHERE

LUMMER—BRODHUN
PHOTOMETER HEAD

Fig. 6.13. Use of an integrating sphere

a photoelectric cell or a Lummer–Brodhun head immediately outside the window (Fig. 6.13). The sphere, once calibrated, can be used for many photometric measurements (for example, for the determination of the m.s.c.p. of a lamp, the reflectance of a mirror or the transmission of a filter).

6.14 The Flicker Photometer

If two luminous sources are to be compared by a photometer, a device such as the Lummer–Brodhun head is not satisfactory if the sources differ considerably in colour because it is difficult for the eye to compare luminances of different colours.

For comparing sources of different colours, a flicker photometer may be used. If two fields of different colour are viewed alternately by the eye and if the alternation is sufficiently rapid, the colour difference is not apparent. Furthermore, when the luminances of the two fields are equal, the flicker disappears. However, increasing the frequency of alternation excessively can

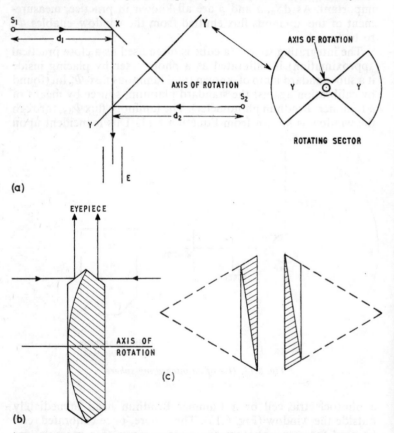

Fig. 6.14. (a) A simple flicker photometer; (b) the Simmance–Abady flicker photometer; (c) construction of the disc of the Simmance–Abady flicker photometer

also cause the flicker to disappear—the period of the alternation becomes shorter than the time of persistence of vision, even though the luminances are unequal. It is, therefore, essential to increase the frequency of alternation slowly from zero until the minimum value is attained at which it is just possible to eliminate the flicker by adjustment to equality of the luminances of the two fields of view.

A simple flicker photometer head [Fig. 6.14(a)] for comparing the luminous intensities of two sources S_1 and S_2 consists of a screen X viewed by an eyepiece E and with a rotating sector Y between X and E. Both X and Y are white diffusing screens formed by coating with magnesium oxide. X is at 45° to light incident from S_1, and Y is perpendicular to X and at 45° to light incident from S_2. As Y is rotated about the axis shown, the field of view observed in the eyepiece is alternately that of X and Y. The speed of rotation of Y is 10–20 rev s^{-1}. The distances d_1 and d_2 are adjusted until no flicker is perceived in the eyepiece, care being taken that the minimum frequency of alternation for this setting is attained. When large colour differences are observed, the speed of Y must be faster.

A useful instrument is the *Simmance–Abady flicker photometer* [Fig. 6.14(b)], where the two screens required are the surfaces of a specially made disc of plaster of Paris. To form this disc, the shaded portions of two truncated cones are removed [as illustrated in Fig. 6.14(c)], and the remaining parts are cemented together at their adjacent faces. When this disc is rotated and its bevelled edges are illuminated and viewed as shown in Fig. 6.14(b), the line dividing these edges swings to and fro across the field of view, providing the alternately illuminated surfaces required.

6.15 Illumination Produced by an Extended Source of Light

Assume that the source of light is a flat circular disc of radius a illuminating the point P on its axis of symmetry. Let the luminance of the disc be L cd cm^{-2}, and let the separation along the axis between P and the centre of the disc be b. Consider an annular ring element of radius r and width dr described about the centre of the disc (Fig. 6.15). This ring element has an area d$A = 2\pi r$ dr, and all points on it are at a distance l from P. l makes an angle θ with the normal at all points on the ring. The luminous intensity dI_θ of this ring in the direction of P is given by

$$dI_\theta = L \, dA \cos \theta = L \, 2\pi r \, dr \cos \theta$$

where L is independent of θ.

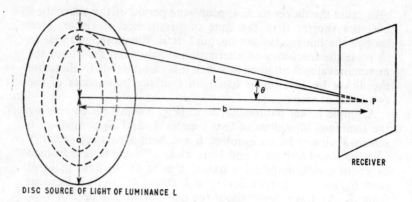

DISC SOURCE OF LIGHT OF LUMINANCE L

Fig. 6.15. *Illumination at a point due to an extended disc source*

The illumination at P due to this annular element is:

$$dE = \frac{dI_\theta \cos \theta}{l^2} = \frac{L \, 2\pi r \, dr \cos^2 \theta}{l^2}$$

Now, $l = b \sec \theta$, $r = b \tan \theta$ and $dr = b \sec^2 \theta \, d\theta$; therefore,

$$dE = \frac{L(2\pi b \tan \theta)b \sec^2 \theta \, d\theta \cos^2 \theta}{b^2 \sec^2 \theta}$$

$$= 2\pi L \sin \theta \cos \theta \, d\theta$$

Hence, the illumination at P due to the entire disc is

$$E = \int_{\theta=0}^{\theta=\alpha} 2\pi L \sin \theta \cos \theta \, d\theta = \pi L \sin^2 \alpha \qquad (6.12)$$

where α is the angle subtended at P by the radius a of the disc. E is given in lumens per square centimetre when L is in candelas per square centimetre.

As $\sin^2 \alpha = a^2/(a^2+b^2)$,

$$E = \frac{L\pi a^2}{a^2+b^2} = \frac{LA}{a^2+b^2} \qquad (6.13)$$

where A is the area of the emitter.

Note that the assumption that the disc is a point source gives the simpler equation $E = LA/b^2$. This introduces a positive error equal to:

$$LA \left(\frac{1}{b^2} - \frac{1}{a^2+b^2} \right) = LA \left[\frac{a^2}{(a^2+b^2)b^2} \right]$$

Therefore, the percentage error is:

$$\frac{LAa^2}{(a^2+b^2)b^2}\times\frac{a^2+b^2}{LA}\times 100\% = \frac{100a^2}{b^2}\%$$

Hence, the error in assuming that the disc is a point source is 1% or less if b is $10a$ or more.

6.16 Measurement of Illumination and Luminance

Measurements with photometer heads, such as the Lummer–Brodhun head, on a photometer bench form standard laboratory practice. Measurements of illumination are frequently required at places other than laboratories, and hence portable apparatus is needed. For example, a desk for comfortable vision should have an illumination of 15 lm ft^{-2}, whilst a drawing table needs twice this amount; in some circumstances (for example, for precision lathe-work or a hospital operating table), an illumination of 200–300 lm ft^{-2} is necessary.

An instrument for measuring illumination of a surface is the *Macbeth illuminometer*. This employs a Lummer–Brodhun photometer cube to compare the illumination of a surface with that of a comparison surface illuminated by an electric lamp within the instrument (Fig. 6.16). In fact, the photometer cube is only able to compare directly luminous intensities from two sources. In the illuminometer, the 'sources' are the illuminated comparison surface and the surface under study. These will emit luminous flux depending on their luminance, not on their illumination. To ensure, therefore, that the illumination of the surface is measured in lumens per square foot, and not its luminance in candelas per square foot, it is essential that both the comparison surface in the instrument and the surface under study be uniformly white diffusing surfaces. This is provided with the instrument: the surface under study (for example, a table top) must be covered with a white diffusing screen during the measurement. This small portable screen is thus an essential part of the instrument.

The distance of the lamp from the comparison surface is varied along a scale in the illuminometer, which may be calibrated directly in units of illumination (for example, lumens per square foot or lumens per square metre). The instrument itself is calibrated by means of observations on a luminous source of known candle power.

The direct measurement of luminance (photometric brightness) of a surface in candelas per unit area is carried out by viewing a

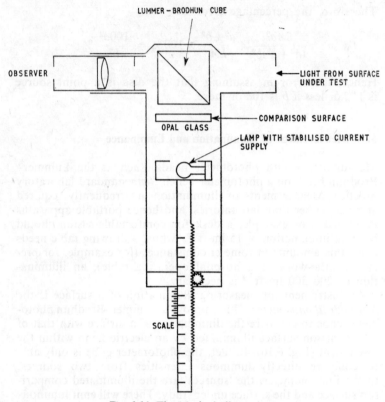

Fig. 6.16. The Macbeth illuminometer

comparison surface S at the same time as the surface X under study (Fig. 6.17). The height of a lamp L in an enclosure is adjusted so that X and S appear to have equal luminance. This height is recorded on a scale Z, which is calibrated (for example, in candelas per square metre) by observation of a standard surface. The distance between the surface X and the entrance aperture to the instrument is immaterial within wide limits because the apparent luminance of an extended source is largely independent of its distance from the observer.

To emphasise the distinction between illumination and luminance, it is noted that a desk is adequately lit for usual reading purposes if the illumination is 15 lm ft^{-2}. On the other hand, if a bright glossy paper were laid on this desk, though the illumination would be the same, the luminance might exceed that comfortably viewed by the eye.

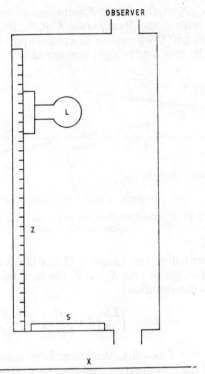

Fig. 6.17. The measurement of luminance of a surface

6.17 The Illumination and Luminance of Optical Images

The illumination of an optical image formed by a lens or a mirror may be calculated if the size, distance and luminance of the object are known.

6.17.1 IMAGE OF AN EXTENDED OBJECT

Suppose the object is an extended source of area S and luminance L and hence is of luminous intensity LS cd. A converging lens of diameter D and area $\alpha = \pi D^2/4$ is used to form an image of this object (Fig. 6.18). If x is the axial separation between the object and the lens, and x' is the axial separation between the image and the lens, the linear magnification is x'/x and the area of the image is $S(x'/x)^2$.

Assuming that the linear dimensions of the object are small compared with x, the illumination E at the lens due to the object is LS/x^2 lm cm^{-2}, where x is in centimetres. If the transmission factor of the lens for the light concerned is T, the total luminous

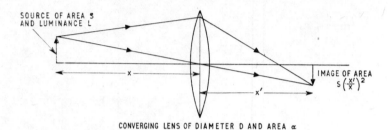

SOURCE OF AREA S
AND LUMINANCE L

IMAGE OF AREA
$S\left(\frac{x'}{x}\right)^2$

CONVERGING LENS OF DIAMETER D AND AREA α

Fig. 6.18. Illumination of an image formed by a converging lens of an extended object

flux transmitted to the image is $(LS/x^2)T\alpha$ lm. As this flux illuminates an image of area $S(x'/x)^2$, the luminance L_I of the image is given by the equation:

$$L_I = \left(\frac{LS}{x^2}\right) T\alpha \frac{x^2}{S(x')^2} = \frac{TL\alpha}{(x')^2} \qquad (6.14)$$

Assuming that T is unity, the image luminance thus depends on the area of the lens aperture and the object luminance, and varies inversely as the square of the image distance.

Putting $T = 1$ and $\alpha = \pi D^2/4$ in Equation 6.14,

$$L_I = \frac{L\pi D^2}{4(x')^2} = \frac{L\pi D^2}{4f^2} \qquad (6.15)$$

for images formed near the second principal focus, where f is the focal length of the lens. As $f/D = N$, the f-number of the lens (Section 1.13),

$$L_I = \frac{L\pi}{4N^2} \qquad (6.16)$$

6.17.2 IMAGE OF A POINT OBJECT

If the object is a point source, as in astronomy, (i.e. if the area S is very small), magnification produces no appreciable increase in size but more light is gathered when an instrument (for example, a telescope) is used than when it is not. From diffraction theory (Section 12.3), the area of the image disc is that area within the first dark ring which subtends an angle $\theta = 1\cdot22$ λ/D, where λ is the wavelength of the light and D is the lens diameter. The radius of the image disc is $f\theta$, where f is the focal length of the lens (for example, the objective lens of a telescope). The area of the image disc is therefore proportional to $(f/D)^2$. The light energy collected by the lens is proportional to LD^2, where L is the luminance of the point source, so the luminance of the image is proportional to LD^4/f^2. Hence, the use of a telescope with an objective of diameter D increases the brightness of the image of a star in proportion to D^4 for a given focal length. Stars which are invisible to the eye can hence be seen with a telescope of sufficiently large aperture.

A more exact treatment of the subject of the brightness of the image formed by an optical instrument requires consideration of the sine relation in geometrical optics (Section 2.5).

An object of small area dS and luminance L has a luminous intensity of LdS per steradian. A coaxial optical system of which the entrance aperture subtends at this axial object a cone of semivertical angle θ will collect luminous flux over a solid angle of $\pi \sin^2 \theta$. The luminous flux radiated from the object within this cone is, therefore,

$$\Phi_v = \pi L \sin^2 \theta \, dS \qquad (6.17)$$

Hence, the luminous flux radiated between angles θ and $\theta + d\theta$ is:

$$d\Phi_v = 2\pi L \sin \theta \cos \theta \, dS \, d\theta \qquad (6.18)$$

This flux emerges from the optical system between angles θ' and $\theta' + d\theta'$, and is given by

$$d\Phi_v' = 2\pi L' \sin \theta' \cos \theta' \, dS' \, d\theta' \qquad (6.19)$$

where L' is the luminance of the image of area dS'. If the transmission of the optical system is T, it also follows that:

$$d\Phi_v' = d\Phi_v T \qquad (6.20)$$

The optical sine theorem states that, for good imaging, each zone of a lens must produce the same magnification. This implies (Section 2.5) that

$$nh \sin \theta = n'h' \sin \theta' \qquad (6.21)$$

where n and n' are the refractive indices of the object and image spaces respectively, whilst h and h' are the heights of the object and image respectively; θ' is the semi-vertical angle of the cone of rays in the image space corresponding to θ in the object space.

Hence,

$$n^2 \, dS \sin^2 \theta = (n')^2 \, dS' \sin^2 \theta' \qquad (6.22)$$

and

$$n^2 \, dS \sin \theta \cos \theta \, d\theta = (n')^2 \, dS' \sin \theta' \cos \theta' \, d\theta' \qquad (6.23)$$

Substituting for $d\Phi'_v$ and $d\Phi_v$ in Equation 6.20 from Equations 6.19 and 6.18 respectively,

$$2\pi L' \sin \theta' \cos \theta' \, dS' \, d\theta' = 2\pi L T \sin \theta \cos \theta \, dS \, d\theta$$

Combining this with Equation 6.23 gives:

$$L' = T \left(\frac{n}{n'}\right)^2 L \qquad (6.24)$$

Usually the object and image are both in air, so $n' = n$ and

$$L' = TL$$

As T is less than unity, it is not possible to make an optical instrument in which the apparant luminance L' of the image exceeds that of the object.

T varies slightly with θ but, neglecting this, the emergent flux from the optical system is

$$\Phi'_v = T\Phi_v = T\pi L \sin^2 \theta \, dS = T\pi L \left(\frac{n'}{n}\right)^2 \sin^2 \theta' \, dS'.$$

from Equations 6.20, 6.17 and 6.22. The illumination of the image is, therefore,

$$E = \frac{\Phi'_v}{dS'} = T\pi L \left(\frac{n'}{n}\right)^2 \sin^2 \theta'.$$

The illumination consequently increases with $\sin^2 \theta'$, so the angle θ' should be as large as possible. This also increases the resolving power (Section 12.2).

6.18 Photoelectric Methods

Photoelectricity may be conveniently defined as the emission of electrons consequent upon the absorption of electromagnetic radiation in a material. There are two main classes of photoelectric phenomena.

The first is concerned with the *photoemissive effect*, or *surface photoemission*, in which electrons are ejected from the surface of a material when radiation is incident upon it. This surface is in the form of a cathode—the *photocathode*—either in a vacuum to permit the electrons released free passage to an anode, or in an inert gas atmosphere to take advantage of gas amplification resulting from ionisation of the gas by the emitted accelerated electrons. The practical devices resulting are the *vacuum photocell* and the *gas-filled photocell*.

The second photoelectric phenomenon occurs when the incident radiation causes electrons to be released within the material and these electrons travel within the material to a collector. Two effects are involved here: *photoconductivity*, which is the increase in the electrical conductivity of a non-metallic solid when exposed to radiation; and the *photovoltaic effect*, in which the absorbed radiation causes an e.m.f. to be set up across a barrier at the discontinuity between, for example, copper and copper oxide, or iron and selenium, or across a p–n junction in a crystalline material such as doped silicon or germanium.

The *photoconductive cell* is the practical development from photoconductivity. The *barrier-layer photocell*, the *photodiode* and the *phototransistor* are the developments from the photovoltaic effect.

The surface photoemission from a pure metallic cathode is in accordance with the Einstein equation

$$hv = \phi + \tfrac{1}{2} m v_{max}^2 \qquad (6.25)$$

where h is Planck's constant $(6 \cdot 625 \times 10^{-27}$ erg s), v is the frequency of the incident radiation, ϕ is the work function energy of the metal, m is the mass of the electron, and v_{max} is the maximum velocity with which the electrons are ejected. The threshold wavelength λ_t at which surface photoemission just takes place is therefore given by

$$\frac{hc}{\lambda_t} = \phi$$

where c is the velocity of light in free space.

For all the metals except the alkali ones (lithium, sodium, potassium, rubidium and caesium), ϕ has a magnitude such that λ_t is in the ultra-violet region of the spectrum. To obtain photoemission in the visible and infra-red regions, special photocathodes are prepared with low values of ϕ. The most useful of these in photometry are: a photocathode prepared by oxidising silver and then activating it with caesium, giving the Cs–O–Ag photocathode; a surface of antimony treated with caesium (the Cs–Sb photocathode); and the caesium–silver–bismuth (Cs–Ag–Bi) photocathode. All three of these exhibit sensitivity to visible light. In addition, two photocathodes sensitive to ultra-violet light are relevant: cuprous iodide and caesium iodide. The former gives maximum photoemission at a wavelength of 1,100 Å, and the latter at 1,250 Å. Cells using these ultra-violet sensitive photo-cathodes would not be used, however, in the photometry of visible light. The chief characteristics of those photocathodes useful in the visible region are given in Table 6.1

Table 6.1. CHARACTERISTICS OF SOME PHOTOCATHODES

Photocathode	λ_m (Å)	λ_t (Å)	Quantum yield
Cs–O–Ag	8,000	12,000	200
Cs–Sb	4,000	7,000	4
Cs–Ag–Bi	4,500	7,500	10

In Table 6.1, λ_m is the wavelength at which the photoemission is a maximum, and λ_t (the threshold wavelength) is taken to be that wavelength at which the photoemission is 1 % of that at λ_m. The quantum yield is the number of photons of wavelength λ_m necessary to cause the ejection of one electron. Thus Cs–Sb is the most efficient of these photocathodes. Cs–O–Ag is one of the earliest photocathode materials, much used in photocells, television pick-up tubes and image converters; it has sensitivity in the near infra-red. The choice for photometric work lies, however, between Cs–Sb and Cs–Ag–Bi, with a preference for the former in view of its efficiency but for the latter in that its peak sensitivity is nearer that of the human eye.

For photometric purposes, the radiant energy has to be meas-ured in terms of its capability of producing a visual response. A photocell used as a sub-standard light-sensitive device in photo-metry will only give correctly the luminous intensity of a source of light if it is calibrated and used under conditions whereby its spectral sensitivity is close to that of the human eye. There is no photocell, of any type, with such a spectral response. The Cs–Sb type has a spectral response curve (Fig. 6.19) with a peak at 4,000 Å; the eye has such a peak at 5,550 Å. To bring the curve for Cs–Sb nearer to that for the eye, an optical filter with appro-priate absorption in the blue region of the spectrum has to be used in front of the photocathode.

A typical commercial vacuum photocell of the Cs–Sb type, with a cathode area of 11·0 cm², gives an output of 45 μA lm^{-1} for light from a lamp of colour temperature 2,700°K. The luminous sensitivity of the vacuum photocell can be enormously enhanced if the photocathode is incorporated within a vacuum tube where

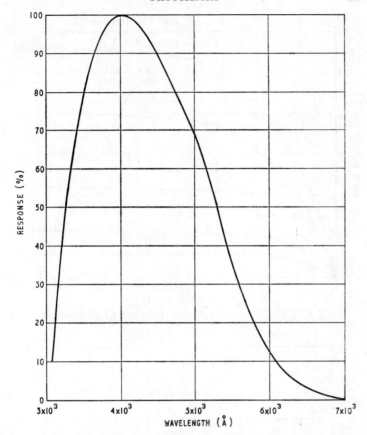

Fig. 6.19. Spectral response of a Cs–Sb photocathode

it is followed by the dynodes of an electron multiplier. A photo-multiplier of this type, with a photocathode area of only 0·8 cm² and nine stages of electron multiplication by secondary emission, can give an output of 2 A lm⁻¹ with a lamp of colour temperature 2,700°K.

The vacuum photocell has the great advantage that the photo-electric current at saturation anode potential is directly propor-tional to the illumination of the photocathode (Fig. 6.20). Sub-sequent increase of the output current by direct current (d.c.) amplification is readily achieved (Fig. 6.21).

The photoconductive cell utilising cadmium sulphide is partic-ularly sensitive to light, gives large variations in current output for small fluctuations in light intensity, and is capable of a

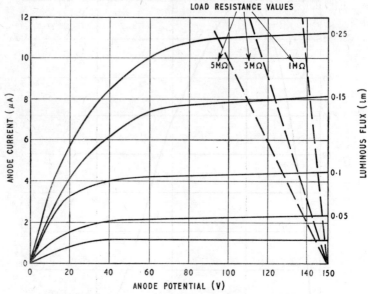

Fig. 6.20. *Current–voltage characteristics of a Cs–Sb photocell*

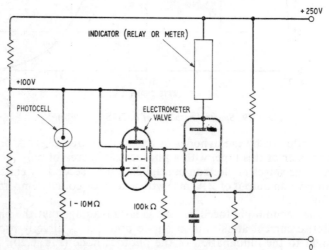

Fig. 6.21. *Basic circuit of a vacuum photocell followed by a d.c. amplifier*

significant output current with a very low level of illumination. The wavelength range in which the response is a maximum is 5,500–7,000 Å. With a sensitive area of about 3 cm², an output current of about 6 mA is obtainable with an illumination of 5 lm ft^{-2} and a d.c. operating e.m.f. of 9 V. This corresponds to a luminous sensitivity of 6 mA per 1/60 lm or 360 mA lm^{-1}.

The cadmium sulphide photoconductive cell is consequently much used in light-operated mechanisms, a well-known example being the modern photoelectric exposure meter for photography, which will operate in moonlight. Unfortunately, changes of sensitivity with life and with the amount of previously incident radiation coupled with indifferent reproducibility of characteristics and a high temperature coefficient make the cadmium sulphide photocell an unsatisfactory choice for accurate photometry.

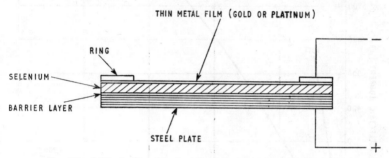

Fig. 6.22. A photovoltaic cell of the barrier-layer type

The photovoltaic cell of the barrier-layer type (Fig. 6.22) is convenient because it does not require a source of e.m.f. for operation. It is connected across a microammeter; the resistance of the microammeter must be such that the relationship between current output and illumination is linear (Fig. 6.23)—enabling a simple photometer to be constructed for which the current is directly proportional to the illumination. The response through the spectrum of an iron–selenium (Fe–Se) cell (Fig. 6.24) has a broad maximum near the wavelength at which the human eye has peak sensitivity but correction by a filter is necessary to give closer agreement.

Photovoltaic cells exhibit fatigue: the current output under steady illumination decreases in the first few minutes of exposure. The decrease depends on the illumination and on the wavelength, being greater in the red end of the spectrum. Their temperature coefficient is of the order of -0.3% per degree Celsius, varying with different cells.

The commercial photovoltaic cell is very convenient for measurements on light because the sensitive surface is plane, of area 5 cm^2 or more and readily accessible, and the mounting of the cell is straightforward. The cell sensitivity depends on the angle of incidence of the light. For a given illumination, the output decreases rapidly for angles of incidence exceeding 60°, owing mainly to the increased reflection of light from the protective glass window or lacquer layer over the sensitive surface. A photovoltaic cell of the selenium type, constantly exposed for three years, showed no changes in characteristics on periodic calibration.

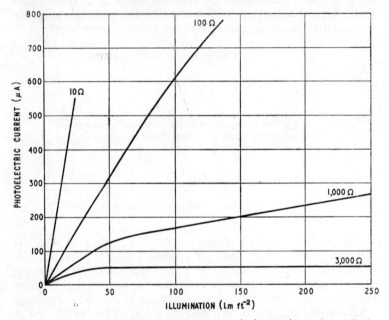

Fig. 6.23. Current–illumination characteristics of a barrier-layer photocell (the parameter is the load resistance across the cell)

For precision photometric work, the vacuum photocell is, in general, preferable to the photoconductive and photovoltaic types.

The silicon photovoltaic cell, the silicon and the germanium photodiodes, and the phototransistor are of great sensitivity to light and are particularly useful when a sensitive surface of small area is needed. The phototransistor, especially, is valuable in such instruments as the microphotometer used for measuring the optical density in a small region of an exposed and developed photographic plate (they are, for example, employed in spectrography). Manufacturers' catalogues provide very good sources of

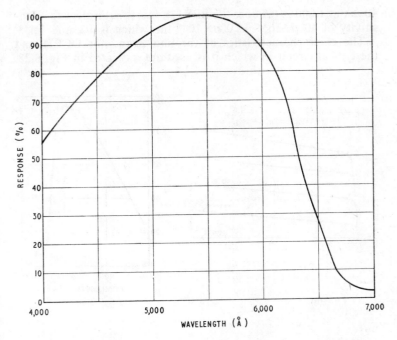

Fig. 6.24. Spectral sensitivity of a typical Fe–Se barrier-layer photocell

information on characteristics. Typical data [based on devices marketed by Mullard Ltd are summarised below.

Silicon photovoltaic cell BPY10: sensitive area, 0.066 in$\times 0.066$ in; sensitivity, 32 μA at 2,000 lx; peak spectral response at 8,000 Å.

Silicon photodiode BPY13: sensitive area, (12 ± 1) mm^2; sensitivity, 40 μA at 1,000 lx; peak spectral response at 9,200 Å.

Germanium photodiode OAP12; sensitive area, 1.0 mm^2; sensitivity, 0.05 μA at 1 lx; peak spectral response in infra-red at 15,500 Å.

Phototransistor OCP71: sensitive area, 7 mm^2; sensitivity, 0.3 A lm^{-1}; peak spectral response at 15,500 A with a cut-off at 20,000 Å.

The lux is an illumination of one lumen per square metre (Section 6.4). A silicon photodiode with a sensitive area of 12 mm^2, therefore, receives a luminous flux of 12×10^{-6} lm when the illumination is 1 lx. Consequently, if the photodiode provides

40 μA at 1,000 lx (as does the BPY13), this is equivalent to a sensitivity of 40 μA per $12,000 \times 10^{-6}$ lm, which is 3·3 mA lm^{-1}.

The chief characteristics of the phototransistor OCP71 and the typical circuit in which it is used are indicated in Fig. 6.25.

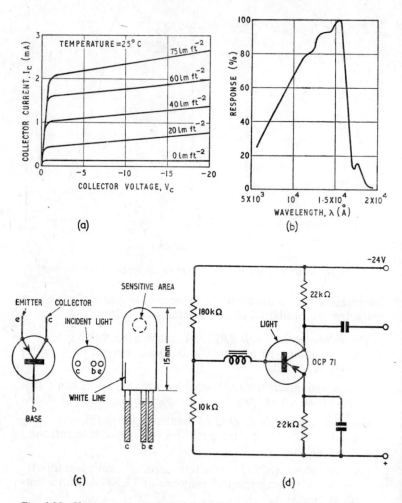

Fig. 6.25. Characteristics and use of the phototransistor OCP71: (a) I_c–V_c characteristics with illumination, in lumens per square foot, as parameter; (b) spectral response; (c) circuit symbol and connections; (d) operating circuit (collector current $I_c = 0·5$ mA)

6.19 Introduction to Illumination Engineering

The eye distinguishes objects by their brightness and colour. The eye naturally seeks bright objects and its aperture (pupil) is involuntarily adjusted to protect the eye against excessive illumination. Efficiency considerations lead to the choice of a spectral distribution of light as close as possible to the spectral distribution of sunlight, thus providing a close match to the colour response of the eye.

The aim of the illumination engineer is to produce adequate illumination of a suitable colour, with the prevention of glare and shadow. Colour may be important for some purposes but not for others. The illumination must be economical both for maintenance and installation. The brightest objects in the field of view determine the sensitivity of the eye, consequently direct lighting is avoided and the aim is to produce a convenient and economical distribution of brightness; direct lighting is best superimposed on a lower level of diffused lighting.

The luminance needed depends on the size of the object, speed of movement, duration of observation, brightness contrast and the colour. Observation of small objects requires high illumination because the size of an object in angular measure (minutes) visible to an observer decreases as the luminance increases. A moving object needs to be highly illuminated if it is to be viewed without eye strain. To minimise fatigue in fine work, good illumination is necessary even though the work can be seen adequately in poor light. Brightness contrast may be helpful if shadows assist in locating objects; this entails the use of direct and diffuse illumination together. For example, in sewing, the shadow of the thread makes it more visible.

Glare is annoying because it distracts attention from work and causes fatigue, which reduces the ability to see. Most light sources are far brighter than is desirable for comfort, so diffusing opal surfaces are necessary.

Shadows are helpful for shape recognition but lighting schemes must avoid shadows which are too dense. The importance of colour is clear from the simple example that red objects appear red because they reflect or scatter red light; if the incident light contains no red radiation, the object will seem dark or even black. Modern fluorescent lighting is possibly the best type for colour matching.

Having considered the requirements for a building, the illumination engineer decides how much lighting is needed having regard to the fittings desired, their spacing and mounting. As far as possible, the lighting power must be efficiently used, and the installations capable of being readily cleaned. The luminous flux emitted by each source and fitting is known, usually from manufacturers' data. The light utilisation factor depends on the colours and areas of the walls and ceiling.

Measurements of luminance are often made with photometers of the photovoltaic cell type, which are calibrated against a standard tungsten lamp source. Tungsten and daylight illumination is continuous through the spectrum but, as the light radiation from mercury or sodium discharge lamps is concentrated in a few discrete wavelengths, a cell correction factor is necessary. Quite wide variations occur but typical correction factors are given for approximate evaluation in Table 6.2. The value as determined by the cell is multiplied by the correction factor to give approximately the effective illumination. It is emphasised that such a table is only a rough guide and ignores the many complex factors involved in precision measurements.

Table 6.2. CORRECTION FACTORS FOR BARRIER-LAYER TYPE PHOTOCELLS

Lamp	Fe–Se cell	Cu–CuO cell
Tungsten	1·00	1·00
Mercury discharge	1·45	1·17
Sodium discharge	1·35	1·43
Fluorescent	1·05	0·95
Daylight	1·00	0·80

Exercise 6

1. Distinguish between *luminous flux*, *luminous intensity*, *illumination* and *brightness*.

   Assuming the cosine law of emission, establish a formula expressing the total light radiated from a plane self-luminous surface of small area in terms of its area and normal brightness. Hence find the emission of a surface the brightness of which is defined as one lambert. (L. P.)

2. Describe an accurate form of photometer head suitable for comparing the luminous intensity of a filament lamp with that of a sub-standard tungsten filament lamp of known luminous intensity.

3. Describe, with an account of the appropriate theory, how an integrating sphere is used to provide a known total luminous flux (in lumens) over the area of its exit window. Explain how this sphere could be used to measure the transmission factor of a small telescope.

4. Derive an expression for the luminous flux emitted by an element of area δS of a perfectly diffusing surface of brightness (luminance) B into a cone of semi-angle θ with its axis perpendicular to the element. Taking account of the relative sizes of the exit pupil of the telescope and the entrance pupil of the eye, compare the retinal illumination produced by an extended object viewed through a telescope with that produced when the same object is viewed by the unaided eye. (L. P.)

5. Write an account, with appropriate diagrams, of a photometer head suitable for comparing the luminous intensities from two sources which differ considerably in colour.

6. Explain, with the help of diagrams, the meanings of the terms *aperture stop, entrance pupil, exit pupil*. Discuss the effect which viewing with an astronomical telescope (consisting of two converging lenses) has upon the apparent luminance (brightness) of an extended object. Show that, for a telescope with a given size of entrance pupil, there is a critical value for the magnifying power M, below which the effect is independent of M. Why can stars be made visible in daylight with a telescope? (L. P.)

7. Lamps are fitted behind a translucent circular plate in the ceiling of a room so that this plate may be regarded as a flat disc-shaped source of light of uniform luminance L stilb with a radius of 30 cm. Calculate the illumination in phot at a point P on the axis of symmetry of this disc at a distance of 100 cm. Evaluate the error in finding this illumination which would result if it were assumed that the light was from a point source.

8. Write an essay on photoelectric devices in which are included accounts of the characteristics and uses in photometry of vacuum photocells, photoconductive cells, barrier-layer photodiodes and phototransistors.

9. Assuming that the Sun is a black-body radiator with a surface temperature of $6,000°K$, show that the total radiation from it is 7.4×10^6 mW cm^{-2}, given that Stefan's constant is 5.67×10^{-12} J cm^{-2} s^{-1} deg^{-4}K.
Sketch a curve showing approximately the distribution of energy in the spectrum of the radiation from the Sun and calculate the wavelength at which the maximum peak occurs in this curve. (The velocity of light = 3×10 cm s^{-1}, Planck's constant = 6.6×10^{-27} erg s and Boltzmann's constant = 1.38×10^{-16} erg deg^{-1}K.)

10. Write a short essay on illumination engineering.

11. Give a critical account of modern methods by which the relative intensities of light sources of different colours may be compared. (G. I. P.)

12. Discuss the relationship between the energy and the luminosity associated with a beam of light. Explain how the relative sensitivity of the eye to different wavelengths differs for high and low levels of illumination.
Discuss the importance of the curve relating energy and luminosity in visual photometry, and explain how it might be obtained experimentally. (G. I. P.)

CHAPTER 7

Wave Theory of Light

Preceding chapters have outlined the principles of those aspects of optical phenomena which can be largely described in terms of geometrical optics, based on the concept that rays travel in straight lines in a homogeneous medium of uniform refractive index free of obstacles. This geometrical approach is limited because there is a whole range of phenomena, including interference, diffraction and polarisation, which cannot be explained by ray concepts. To discuss adequately these phenomena, the wave nature of light must be considered. The geometrical theory is also limited with regard to image formation by a lens or mirror system because ray concepts only give the macroscopic picture; it is left to wave theory to give a more detailed picture on the microscopic scale, so to speak, leading to the concept of optical resolution.

Certain fundamental aspects of wave theory apply to waves in material media, such as waves on liquids, vibrations in solids and the propagation of sound. Here, the wave motion is in an elastic medium, and displacements of material particles (for example, of gas molecules when sound is propagated in a gas) are involved. With light and the other radiations in the electromagnetic spectrum, a material medium is not concerned: the propagation can take place in free space. The electromagnetic theory of radiation is now involved. This will be considered more fully in Chapter 1 of Volume 2. For the present, it suffices to state that light is propagated as an electromagnetic wave in which the electric and magnetic field vectors are perpendicular to the direction of propagation; the wave motion is transverse. In an elastic medium, the displacement of a particle at a particular time is involved;

for light, however, the magnitude of the electric or magnetic vector at a particular time is the concern. Frequently, attention can be conveniently confined to the electric vector alone. For the propagation of electromagnetic radiation (which includes visible light), a displacement y transverse to the direction of propagation x is then thought of as the magnitude of an electric vector.

The wave theory in its many aspects was predominant in optics until the advent of the quantum theory; this involves the concept of the photon necessary to explain phenomena, such as the photoelectric effect and atomic spectra, which are inexplicable on a basis of electromagnetic theory. In terms of quantum theory, in which the photon has an energy of hf where h is Planck's constant and f is the frequency, the idea of a wave appears inexact and inadequate. Wave theory is, however, sufficiently exact to explain satisfactorily all phenomena in radiation except those involving mechanisms on the atomic and molecular scale of dimensions—in particular, wave theory provides satisfactory explanations of the production of radiation and the interactions between radiation and matter.

7.1 The Equations Representing Progressive Waves

Fundamentally, electromagnetic radiation is brought about by the acceleration of electric charge. The radiation from a heated body, such as an electric lamp filament, is due to the motion of the molecules or atoms, containing electrons, within the hot material; the radiation from a gaseous discharge is caused by the motion of electrons in the excited gas atoms or molecules. Quantum theory is essential to explain the production of such radiation. The result in the surrounding space is, nevertheless, electromagnetic radiation—which, if it is within the wavelength range between 3,800 Å and 7,600 Å approximately, is visible light. This radiation is emitted continuously as long as the source is excited. In the surrounding space, the effect is a continuous train of waves which are periodic but, in general, of complex wave form.

At any point a given distance from the source, the displacement y (for instance, the magnitude of the electric vector in volts per centimetre) will vary periodically with time as the wave passes this point, and radiant energy will flow past this point. At a different distance from the source, at the same instant of time, the displacement y will, in general, be different. Along an x-axis, therefore, at a given instant of time, the profile of the 'frozen' wave motion is represented in general by

$$y = \phi(x - x_0)$$

where ϕ is some periodic function, and y is the transverse displacement at the distance x from a suitable origin taken to be at x_0.

A periodic wave of this kind is represented by the full-line waveform of Fig. 7.1. This is a 'snapshot' of the wave at some instant of time t; the wave is frozen.

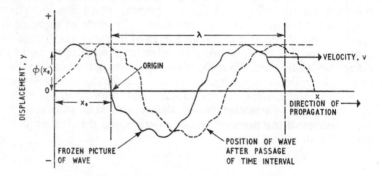

Fig. 7.1. A periodic wave

After the passage of an additional short time interval, the new position of the wave is represented in Fig. 7.1 by the dotted line. If this time interval is t_1, the wave will progress a distance vt_1; i.e. some singularity in it corresponding to a particular value y_1 of the displacement y will move a distance vt_1, where v is the velocity with which the wave is progressing. If the wave profile is moving towards the positive direction of x, a particular value of $y = y_1$ which occurred at $x = x_1$ must be repeated at $x = x_1 + vt_1$. The effect will be the same as moving the arbitrary origin through a distance vt_1 in the opposite direction. The expression for the progressive wave hence becomes:

$$y = \phi[(x - x_0) - vt]$$

This is seen to be correct because substitution of $x = x_1$ at $t = 0$ gives the same displacement y_1 as substitution of $x = x_1 + vt_1$ at $t = t_1$.

The value of x_0 depends on the choice of the origin of x at $t = 0$. Generally, this is taken to be zero. The equation for a progressive wave then becomes

$$y = \phi(x \mp vt) \tag{7.1}$$

where the minus sign applies if the wave is progressing in the positive direction of x, and the positive sign is concerned if the progression is in the negative direction of x.

In the ideal case of a train of simple harmonic monochromatic waves having a plane wavefront and progressing in an isotropic medium without dispersion and without attenuation, the periodic function involved is the sine (or cosine) of an appropriate angle. If the maximum displacement (peak value) in the wave is A, it will be repeated at angle intervals of 2π rad corresponding to a change of distance equal to the wavelength λ or an integral multiple of λ. The distance $x-vt$ in Equation 7.1 for such a plane progressive sinusoidal wave travelling in the positive direction of x corresponds to a change of angle of $(2\pi/\lambda)(x-vt)$ and

$$y = A \sin \frac{2\pi}{\lambda}(x-vt) = A \sin 2\pi \left(\frac{x}{\lambda} - \frac{t}{T}\right) \qquad (7.2)$$

because $v/\lambda = T$, the period of the wave motion.

This Equation 7.2 may be expressed in an alternative form because $2\pi/T = \omega$, the angular velocity or pulsatance. Thus,

$$y = A \sin \left(\frac{2\pi x}{\lambda} - \omega t\right) = -A \sin \left(\omega t - \frac{2\pi x}{\lambda}\right)$$

As the change of sign of A and a change from sine to cosine corresponds to a change of origin, so a progressive plane unattenuated sinusoidal wave can be expressed as:

$$y = A \cos \left(\omega t \pm \frac{2\pi x}{\lambda}\right) \qquad (7.3)$$

This displacement y may also be represented by the projection of a rotating vector which has a length equal to the amplitude A of the wave and an angular velocity of rotation of ω. Such a rotating vector may also be constructed in an Argand diagram (Fig. 7.2). As

$$A \exp j\theta = A(\cos \theta + j \sin \theta)$$

where $j = \sqrt{(-1)}$,

$$y = A \exp j \left(\omega t \pm \frac{2\pi x}{\lambda}\right) \qquad (7.4)$$

where the real part of this expression represents the displacement y. Equation 7.4 forms a most convenient alternative to Equation 7.3.

So far, the plane progressive wave in one dimension only has been considered. In general, the normal to the plane wavefront can be represented by a vector n orientated in any direction with

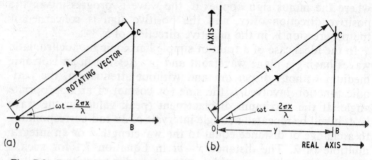

Fig. 7.2. *Rotating vector representation of a plane unattenuated progressive cosine wave: (a) projection $OB = y = A \cos(\omega t - 2\pi x/\lambda)$; (b) corresponding Argand diagram in which the rotating vector OC is the complex quantity $A \exp j(\omega t - 2\pi x/\lambda)$*

respect to three mutually perpendicular Cartesian axes OX, OY, OZ or to a given vector r. The equation of the wavefront will then be $rn \cos \theta = $ constant, or the scalar product of r and n is a constant, i.e.

$$r \cdot n = \text{constant}$$

Thus, for an unattenuated sinusoidal wave, Equation 7.4 becomes

$$\psi = A \exp j(\omega t - kr \cdot n) \tag{7.5}$$

where $k = 2\pi\lambda$, and ψ replaces y because the displacement is perpendicular to n which is not, in general, along the x-axis.

In terms of the direction-cosines $l = \cos \alpha$, $m = \cos \beta$ and $n = \cos \gamma$ of the normal n to the wavefront (Fig. 7.3),

$$r \cdot n = xl + ym + zn$$

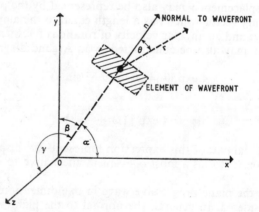

Fig. 7.3. *Plane wave front referred to Cartesian axes*

and

$$\psi = A \; \exp \; j[\omega t - k(lx + my + nz)] \tag{7.6}$$

If this equation is differentiated partially twice with respect to time and twice with respect to x, y and z in succession, it is seen that

$$\frac{\partial^2 \psi}{\partial x^2} + \frac{\partial^2 \psi}{\partial y^2} + \frac{\partial^2 \psi}{\partial z^2} = \frac{1}{v^2} \frac{\partial^2 \psi}{\partial t^2}$$

or

$$\nabla^2 \psi = \frac{1}{v^2} \frac{\partial^2 \psi}{\partial t^2} \tag{7.7}$$

where ∇^2 is the Laplacian operator expressed in Cartesian co-ordinates as:

$$\nabla^2 \equiv \left(\frac{\partial^2}{\partial x^2} + \frac{\partial^2}{\partial y^2} + \frac{\partial^2}{\partial z^2} \right)$$

Equation 7.7 is the differential equation of an undamped sinusoidal wave of phase velocity v.

7.2 Progressive Spherical and Cylindrical Waves

A point source in an isotropic medium will emit spherical waves. The wavefront at a distance r from the source will be a sphere of radius r. If the source is monochromatic, the displacements at a given time at all points in this wavefront will be in phase. (The following treatment also applies to spherical waves converging to a point image.)

For waves of a single frequency f, pulsatance ω and wavelength λ, the displacement ψ at a time t is given by

$$\psi = \frac{A}{r} \cos (\omega t - kr) \tag{7.8}$$

where A is the amplitude at unit distance from the source when $r = 1$ and $k = 2\pi/\lambda$. As the rate of flow of energy through unit area perpendicular to the direction of propagation is the intensity, which is proportional to the square of the amplitude, it follows that the intensity at a distance r from the source depends on $(A/r)^2$, so falls off inversely with r^2; this is in accordance with the well-known inverse square law for radiant energy.

The differential equation for spherical waves is best reached by the use of vector calculus to avoid cumbersome mathematical expressions. Suffice to state for the present that the appropriate differential equation is

$$\frac{1}{v^2}\frac{\partial^2\psi}{\partial t^2} = \frac{\partial^2\psi}{\partial r^2} + \frac{2}{r}\frac{\partial\psi}{\partial r} \tag{7.9}$$

of which the general solution is

$$r\psi = \phi_1(\omega t - kr) + \phi_2(\omega t + kr) \tag{7.10}$$

This represents one spherical wave of pulsatance ω emanating from a point O and one converging towards O, where k is a constant. A particular harmonic solution of Equation 7.10 is

$$\psi = \frac{A}{r}\exp j(\omega t - kr) \tag{7.11}$$

of which the real part is the same as Equation 7.8.

For a cylindrical wavefront emanating from an infinitely thin line source or converging towards a line image, the differential equation is

$$\frac{1}{v^2}\frac{\partial^2\psi}{\partial t^2} = \frac{\partial^2\psi}{\partial \varrho^2} + \frac{1}{\varrho}\frac{\partial\psi}{\partial \varrho} \tag{7.12}$$

where ϱ is the radius of the cylindrical wave surface. A particular harmonic solution is

$$\psi = \frac{A}{\sqrt{\varrho}}\exp j(\omega t - k\varrho) \tag{7.13}$$

where A is the amplitude at unit distance from the source.

Note that in a spherical wave, the amplitude at a distance r from the source is proportional to $1/r$, and the intensity is proportional to $1/r^2$; on the other hand, in a cylindrical wave, the amplitude is proportional to $1/\sqrt{\varrho}$, where ϱ is the radius of the cylindrical surface, and the intensity is proportional to $1/\varrho$.

7.3 The Principle of Superposition of Wave Motions

The propagation of any one beam of radiation is unaffected by the presence of other beams. If a beam of light crosses a second beam, superposition will occur in the region of intersection but the one beam will subsequently proceed as it would have done if the other had not been present.

In the region where superposition occurs, two or more progressive waves act on the same elastic particle at any point or, for electromagnetic radiation, the waves produce electric field vectors in the same region around any point. The resultant displacement at any instant is the vector sum of the separate displacements due to the individual waves. If the state of polarisation of the superposed waves is the same at any instant (which, for light, means that the directions of the electric vectors in the waves are the same at any point at a given instant), the vector sum becomes simply the scalar sum of the separate displacements.

This principle of superposition is fundamental to the theory of interference and diffraction; it enables calculations to be made of the amplitude, and so the intensity, at any point due to waves from sources which interfere. Interference can, however, only be observed if the two (or more) sources of the wave systems radiate in phase, i.e. are *coherent*. It is only then that a stationary interference pattern will be observed. If the two sources result in waves of which the phase and/or wavelength are continually changing, the interference pattern will never be stationary and will remain unobserved. With interference of light, it is never possible to obtain two (or more) coherent waves systems from sources unless they are derived from one single source. Any change in the master source then affects both equally, and the interference pattern is stationary.

The simplest example of superposition is that of two simple linear sine waves of the same frequency in the same state of polarisation in an isotropic medium without dispersion and arriving at a region where they are superposed with different phases. The separate waves may be represented by

$$y_1 = A_1 \sin\left(\omega t - \frac{2\pi x}{\lambda} + \alpha_1\right)$$

and

$$y_2 = A_2 \sin\left(\omega t - \frac{2\pi x}{\lambda} + \alpha_2\right)$$

where one wave has an amplitude A_1, and the other A_2. The two respective phases at some point in the region of superposition are represented by α_1 and α_2. As the resultant displacement y due to the superposed waves is required in a region specified by a certain value of x, x will be a constant and can be included as a phase. The equations can hence be conveniently written in the form

$$y_1 = A_1 \sin(\omega t - \delta_1)$$

and

$$y_2 = A_2 \sin(\omega t - \delta_2)$$

where δ_1 and δ_2 are the phase constants.

The resultant disturbance is given by:

$$y = y_1 + y_2$$
$$= A_1 \sin (\omega t - \delta_1) + A_2 \sin (\omega t - \delta_2)$$
$$= (A_1 \cos \delta_1 + A_2 \cos \delta_2) \sin \omega t - (A_1 \sin \delta_1 + A_2 \sin \delta_2) \cos \omega t$$

Let

$$A_1 \cos \delta_1 + A_2 \cos \delta_2 = R \cos \delta$$

and

$$A_1 \sin \delta_1 + A_2 \sin \delta_2 = R \sin \delta$$

Then

$$y = R \cos \delta \sin \omega t - R \sin \delta \cos \omega t = R \sin(\omega t - \delta) \quad (7.14)$$

where

$$R^2 = A_1^2 + A_2^2 + 2A_1 A_2 \cos (\delta_2 - \delta_1) \quad (7.15)$$

and

$$\tan \delta = \frac{A_1 \sin \delta_1 + A_2 \sin \delta_2}{A_1 \cos \delta_1 + A_2 \cos \delta_2} \quad (7.16)$$

If the phase difference $\delta_2 - \delta_1$ between the two linear sinusoidal waves is zero or an even multiple of π, corresponding to an optical path difference of zero or $(\lambda/2\pi)(\delta_2 - \delta_1) = n\lambda$ where n is an integer, $\cos (\delta_2 - \delta_1)$ is $+1$ and Equation 7.15 becomes:

$$R^2 = A_1^2 + A_2^2 + 2A_1 A_2 = (A_1 + A_2)^2$$

Therefore,

$$R = A_1 + A_2$$

i.e. the waves are in phase, so their amplitudes are directly additive, giving *constructive interference*.

If the phase difference $\delta_2 - \delta_1$ is an odd multiple of π, corresponding to the optical path difference being an odd multiple of $\lambda/2$, $\cos (\delta_2 - \delta_1)$ is -1 and Equation 7.15 becomes:

$$R^2 = A_1^2 + A_2^2 - 2A_1 A_2 = (A_1 - A_2)^2$$

Therefore,

$$R = A_1 - A_2$$

i.e. the waves are in anti-phase, so their amplitudes are directly subtractive, giving *destructive interference*.

If $A_1 = A_2$, constructive interference results in an amplitude twice that due to the individual waves, and the resultant intensity is four times that due to one wave. On the other hand, destructive interference results in the amplitude and intensity both being zero. If $A_1 = A_2 = A$ and a phase difference of $\delta_2 - \delta_1$ exists, it can be seen from Equation 7.15 that the resultant amplitude is given by:

$$R = \sqrt{\{2A^2[1 + \cos(\delta_2 - \delta_1)]\}} \qquad (7.17)$$

The resultant intensity I, proportional to the square of the amplitude R, is therefore given by

$$I = k\ 2A^2[1 + \cos(\delta_2 - \delta_1)] = 4A^2k \cos^2[(\delta_2 - \delta_1)/2] \quad (7.18)$$

where the multiplication factor k involved depends upon the units in which A and I are specified. In general, k is taken to be unity for convenience in relating I to R^2 but it will not be unity if R is measured in, say, centimetres and photometric units are employed for I.

If the linear sine waves in the same state of polarisation which are superposed have various frequencies and corresponding pulsatances $\omega_1, \omega_2, \omega_3, \ldots$ and amplitudes $A_1, A_2, A_3, \ldots$ respectively, the various displacements $y_1, y_2, y_3, \ldots$ at a time t are represented by

$$y_1 = A_1 \sin\left(\omega_1 t - \frac{2\pi x_1}{\lambda_1}\right)$$

$$y_2 = A_2 \sin\left(\omega_2 t - \frac{2\pi x_2}{\lambda_2}\right)$$

and so on, where $\lambda_1, \lambda_2, \lambda_3, \ldots$ are the respective wavelengths. Each of these progressive waves may be represented by a rotating vector where successive vectors are OA_1 of length A_1, A_1A_2 of length A_2, and so on; the resultant y is the sum of the separate projections in a given direction at a given instant of time, as shown in Fig. 7.4.

When the pulsatances of the sine waves are all the same, i.e. when $\omega_1 = \omega_2 = \omega_3 = \ldots$, the vectors $OA_1, A_1A_2, A_2A_3, \ldots$ will all rotate with the same angular velocity, and this will also be the velocity of the resultant vector OA_N. The polygon as a whole rotates, and the peak value of the resultant displacement is represented by vector OA_N. If, however, the waves have different pulsatances, the shape of the polygon changes with time, so OA_N will be a function of time as well as of the spatial coordinates. This state of affairs corresponds to amplitude or peak variation with time, i.e. *amplitude modulation*. The shape of the polygon will also change if the spatial parts of the phase angles change; this often arises in problems of interference and diffraction.

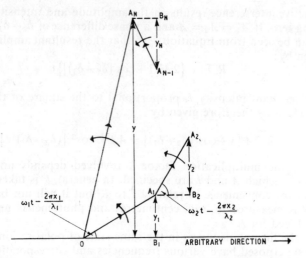

Fig. 7.4. *Vector addition of superimposed progressive simple linear sine waves*

In the analytical treatment of the superposition of a number of wave motions, the use of complex quantities is convenient. Since

$$\exp j \, x = \cos x + j \sin x$$

an expression of the form

$$y = A \cos (\omega t - \delta)$$

can be considered as the real part of the complex quantity

$$y = A \exp j(\omega t - \delta)$$

As physical significance can only be related to the real part, the extraction of the real part of the sum of a number of such complex quantities representing individual waves will give the resultant. Again, instead of representing the displacement in the wave motion by $A \exp j(\omega t - \delta)$ it can be expressed as

$$y = A \exp j(\omega t - \delta) = A \exp (-j\delta) \exp j\omega t$$

consisting of a complex amplitude $A \exp (-j\delta)$, and a harmonic variation with time represented by $\exp j\omega t$.

Using this procedure, consider again the superposition of two linear sinusoidal waves in the same state of polarisation, represented by

$$y_1 = A_1 \exp j(\omega_1 t - \delta_1) = A_1 \exp (-j\delta_1) \exp j\omega_1 t$$

and
$$y_2 = A_2 \exp j(\omega_2 t - \delta_2) = A_2 \exp(-j\delta_2) \exp j\omega_2 t$$

If the pulsatances ω_1 and ω_2 are the same and each is equal to ω, the result of superposition is represented by:

$$y = [A_1 \exp(-j\delta_1) + A_2 \exp(-j\delta_2)] \exp j\omega t$$

The complex amplitude of this expression is:

$$z = A_1 \exp(-j\delta_1) + A_2 \exp(-j\delta_2)$$

The complex conjugate of z is

$$\bar{z} = A_1 \exp j\delta_1 + A_2 \exp j\delta_2$$

and $z\bar{z} = |z|^2$ where $|z|$ is the modulus of z. The intensity is, therefore, represented by

$$
\begin{aligned}
I &= R^2 \\
&= z\bar{z} \\
&= [A_1 \exp(-j\delta_1) + A_2 \exp(-j\delta_2)](A_1 \exp j\delta_1 + A_2 \exp j\delta_2) \\
&= A_1^2 + A_2^2 + A_1 A_2 [\exp j(\delta_1 - \delta_2) + \exp(-j)(\delta_1 - \delta_2)] \\
&= A_1^2 + A_2^2 + 2A_1 A_2 \cos(\delta_1 - \delta_2)
\end{aligned}
$$

which is the same as Equation 7.15 because $\cos(\delta_1 - \delta_2) = \cos(\delta_2 - \delta_1)$.

7.4 Superposition of Simple Harmonic Waves of Slightly Different Frequencies and Velocities: Group and Wave Velocity

If the wave motion takes place in a dispersive medium, the velocity of the wave changes with wavelength because of the variation of refractive index with wavelength. In Section 7.3, this dispersion has been ignored. If a propagated group of waves traverses a dispersive medium and the source of light contains components of various wavelengths, the component waves will not all have the same velocity.

To simplify the treatment, suppose there are only two component simple harmonic waves traversing the same path and consequently superposed, where wave 1 has a velocity v and wavelength λ, and the other, wave 2, has velocity $v + \delta v$ and wavelength $\lambda + \delta\lambda$. Let the amplitudes of the two waves be the same and equal to A. The displacement y_1 of wave 1 is given by an expression of the form of Equation 7.2:

$$y_1 = A \sin \frac{2\pi}{\lambda}(x - vt)$$

The displacement y_2 in wave 2 is represented by:

$$y_2 = A \sin \frac{2\pi}{\lambda + \delta\lambda}[x - (v + \delta v)t]$$

The superposition results in a displacement y given by the equation:

$$y = y_1 + y_2 = A\left\{\sin \frac{2\pi}{\lambda}(x - vt) + \sin \frac{2\pi}{\lambda + \delta\lambda}[x - (v + \delta v)t]\right\}$$

Therefore,

$$y = 2A \sin \pi \left[x\left(\frac{1}{\lambda} + \frac{1}{\lambda + \delta\lambda}\right) - t\left(\frac{v}{\lambda} + \frac{v + \delta v}{\lambda + \delta\lambda}\right)\right] \times$$

$$\cos \pi \left[x\left(\frac{1}{\lambda} - \frac{1}{\lambda + \delta\lambda}\right) - t\left(\frac{v}{\lambda} - \frac{v + \delta v}{\lambda + \delta\lambda}\right)\right]$$

If $\delta\lambda$ is small compared with λ,

$$\frac{1}{\lambda} - \frac{1}{\lambda + \delta\lambda} = \frac{\delta\lambda}{\lambda^2 + \lambda\delta\lambda} = \frac{\delta\lambda}{\lambda^2}$$

$$\frac{1}{\lambda} + \frac{1}{\lambda + \delta\lambda} = \frac{2\lambda + \delta\lambda}{\lambda^2 + \lambda\delta\lambda} = \frac{2}{\lambda}$$

$$\frac{v}{\lambda} - \frac{v + \delta v}{\lambda + \delta\lambda} = \frac{v\delta\lambda - \lambda\delta v}{\lambda^2}$$

and

$$\frac{v}{\lambda} + \frac{v + \delta v}{\lambda + \delta\lambda} = \frac{2v\lambda + v\delta\lambda + \lambda\delta v}{\lambda^2} = \frac{2v}{\lambda}$$

Therefore,

$$y = 2A \sin \frac{2\pi}{\lambda}(x - vt) \cos \frac{\pi\delta\lambda}{\lambda^2}\left[x - t\left(\frac{v\delta\lambda - \lambda\delta v}{\delta\lambda}\right)\right]$$

In this expression, the term $2A \sin [(2\pi/\lambda)(x - vt)]$ represents a wave propagated in the positive direction of x with a speed v and wavelength λ, whereas the term $\cos \{(\pi\delta\lambda/\lambda^2)[x - t(v\delta\lambda - \lambda\delta v)/\delta\lambda]\}$ represents a wavelength of $2\lambda^2/\delta\lambda$—which is much longer than λ—propagated in the positive direction of x with a speed $(v\delta\lambda - \lambda\delta v)/\delta\lambda$. The first term is plotted as the full line in Fig. 7.5, and the second term is the comparatively slowly varying envelope plotted as the dotted line.

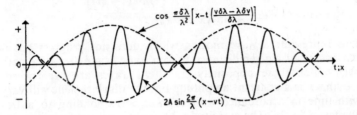

$$\cos \frac{\pi\delta\lambda}{\lambda^2}\left[x-t\left(\frac{v\delta\lambda - \lambda\delta v}{\delta\lambda}\right)\right]$$

$$2A \sin \frac{2\pi}{\lambda}(x-vt)$$

Fig. 7.5. Group and wave velocity

The wave velocity is v, whereas the velocity of the group is v_g, where

$$v_g = \frac{v\delta\lambda - \lambda\delta v}{\delta\lambda} = v - \lambda\frac{\delta v}{\delta\lambda}$$

Therefore, the equation

$$v_g = v - \lambda\frac{dv}{d\lambda} \tag{7.19}$$

gives the group velocity in terms of the wave velocity and the dispersion of the medium (which decides $dv/d\lambda$). Equation 7.19 is the same as Equation 5.7 obtained by a simpler analytical method.

7.5 Superposition of Simple Harmonic Waves of the Same Frequency Travelling in Opposite Directions: Stationary Waves

If two simple harmonic waves of the same frequency are propagated in a non-dispersive medium, one being propagated with the speed v in the positive direction of x and the other with the same speed in the negative direction of x, they can be represented by

$$y_1 = A \sin k(x-vt)$$

and

$$y_2 = A \sin k(x+vt)$$

where the amplitudes are the same.

If the waves are polarised in the same direction, then the displacement y in the resultant is given by:

$$y = y_1 + y_2$$
$$= A[\sin k(x-vt) + \sin k(x+vt)]$$
$$= 2A \sin kx \cos kvt$$

This represents a wave motion which does not progress but is *stationary*. At a given distance x from some origin along the path over which these oppositely travelling waves are superposed, $2A \sin kx$ is a constant amplitude but the displacement will vary with time in accordance with $\cos kvt$ corresponding to a frequency of $kv/2\pi$. The amplitude $2A \sin kx$ will attain maximum values when $\sin kx = 1$, i.e. when $kx = (2m+1)(\pi/2)$, where m is zero or an integer. These maximum amplitudes in the wave occur at the anti-nodes, which are separated by distances corresponding to a change of m by unity. Therefore, the separation between adjacent anti-nodes at positions $x = x_1$ and $x = x_2$ is given by:

$$k(x_1 - x_2) = [2(m+1)+1]\frac{\pi}{2} - (2m+1)\frac{\pi}{2} = \pi$$

Therefore,

$$x_1 - x_2 = \frac{\pi}{k}$$

But $k = 2\pi/\lambda$, where λ is the wavelength of either wave. Hence the separation between adjacent anti-nodes is:

$$\frac{\pi}{k} = \frac{\pi}{2\pi/\lambda} = \frac{\lambda}{2}$$

The amplitudes in the displacement will be zero at values of x decided by $\sin kx = 0$, i.e. when $kx = m\pi$, where m is zero or an integer. At these zeros or *nodes* $x = m\pi/k$, and the separation between adjacent nodes is $\pi/k = \lambda/2$; they will occur midway between the anti-nodes.

7.6 Huygens' Principle

In 1678 (published in 1690), Huygens evolved the wave theory of light and, in particular, enunciated the famous principle that every point on a wavefront may be considered as the centre of a disturbance, in fact, as a source of *secondary wavelets*. The enve-

lope of these secondary wavelets gives the wavefront at any sub-
sequent time. For example, let AB in Fig. 7.6 be, at a particular
instant of time, a section of a spherical wavefront 1 which ema-
nates from a point source S; the position of the wavefront 2 at
a time t later is found by taking every point in succession in AB
as a centre and constructing spheres of radii vt about these
points, where v is the velocity of light in the isotropic uniform
medium. These spheres represent secondary wavelets and have
an envelope $A'B'$, which is the location of the later wavefront 2.

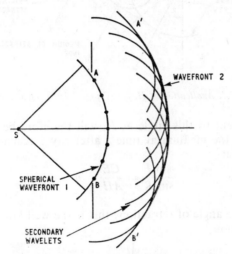

Fig. 7.6. An application of Huygens' principle

In its simplest form, this principle may be used to verify the
laws of reflection and refraction and to predict approximately
field diffraction effects. The application to refraction at a plane
boundary separating two different uniform media is illustrated by
Fig. 7.7. A collimated beam of monochromatic light has extreme
rays 1 and 2, and is incident on the boundary at an angle I.
AC is a plane wavefront in this incident beam when ray 1 has
reached A at the boundary but ray 2 has still to traverse the distance CE before reaching the boundary; AC is perpendicular
to ray 2 at C. The distance $CE = ct$, where t is the time taken for
ray 2 with velocity c *in vacuo* (i.e. in air taken to have refractive
index of unity) to traverse the distance CE. During this time, ray
1 will be propagated with velocity c/n in the medium of refractive index n. It will reach H, where $AH = ct/n = CE/n$. An arc
of a circle of radius CE/n is therefore drawn about A as centre.

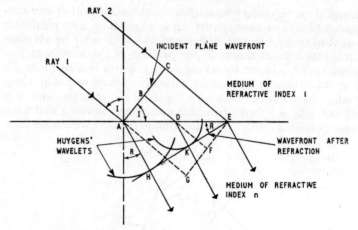

Fig. 7.7. Application of Huygens' principle to refraction

EH, the tangent to this circle at *H*, will therefore be the plane wavefront in the medium at time *t* after ray 1 reaches *A*. It is easily seen that

$$\frac{\sin I}{\sin R} = \frac{CE}{AH} = n$$

where *R* is the angle of refraction; this is the well-known Snell's law of refraction.

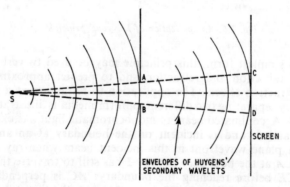

Fig. 7.8. Illumination of a small aperture by a point source of light

Fig 7.8 illustrates spherical waves from a point source of light *S* which are incident upon a small aperture *AB* in a shield. The Huygens' secondary wavelets spread out from points in the wavefront at the aperture. A screen placed beyond the aperture on the side opposite to that of the source would, in accordance with the principle of the rectilinear propagation of light, be illuminated within the patch CD, where SAC and SBD are straight lines. Outside this patch there would be shadow, with a sharp edge at CD. The Huygens' secondary wavelets show that this is not the result: the secondary wavelets spread beyond the cone defined by AC and BD; light does appear in the shadow, particularly near the geometrically defined edge. Thus, the shadow edge is not sharp but subject to alternations of light and shadow owing to the manner in which the secondary wavelets interfere with one another. This is the phenomenon known as *diffraction*.

This diffraction is present irrespective of the size of the aperture, but the larger the aperture is in relation to the wavelength of the radiation the less significant is the effect of diffraction. If the aperture in Fig. 7.8 is a circular hole of diameter 5 cm, say, the diffraction effects at the edge of the shadow will be of minor importance; for most practical purposes, the edge will be sharp and determined by the assumption of rectilinear propagation. On the other hand, a hole of diameter 1 mm, say, would result in diffraction effects predominant in determining the spread of light beyond the aperture and hence the alternations of light and shade at the screen; the assumption of rectilinear propagation would then give totally misleading ideas about the nature of the shadow.

The method involving the construction of Huygens' wavelets is useful but limited: it does not explain why the wavelet considered to emanate from a point in a wavefront is not propagated backwards as well as forwards; no prediction is made about the phase of the disturbance due to a wavelet, and the variation of the amplitude of the disturbance with obliquity. The Huygens' procedure was extended by Fresnel and later in the mathematical analysis of Kirchhoff to take these factors into account.

7.7 Change of Phase on Reflection at a Boundary Separating Two Optical Media

A collimated narrow beam 1 of monochromatic light of wave amplitude A in air is incident upon a plane boundary of a transparent medium of refractive index n relative to air [Fig. 7.9(a)]. The reflected beam 2 consists of waves of amplitude Ar, where r is the fraction reflected, whilst the beam 3 refracted into the medium has an amplitude At, where t is the fraction transmitted.

Following the argument put forward by Stokes, suppose the

reflected and refracted beams are reversed in direction. A wave motion is strictly reversible providing there is no loss of energy due to absorption in the medium. Therefore, as shown in Fig. 7.9(b), the reversal of reflected ray 2 of amplitude Ar will give rise to a reflected wave of amplitude Ar^2 along the path of beam 1 (but in the opposite direction), and a transmitted wave of amplitude Art along path 4. The reversal of the refracted beam 3 of amplitude At produces along path 4 a reflected wave of amplitude Atr' (where r' is the fraction reflected at the boundary when incidence is in the medium of refractive index n), and a transmitted wave along the path of beam 1 of amplitude Att' (where t' is the fraction transmitted on passage from the medium of refractive index n to air).

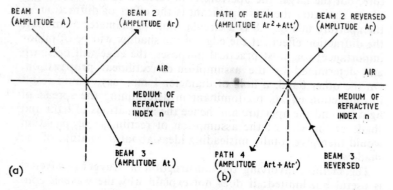

Fig. 7.9. (a) Reflection and refraction at a boundary between two optical media; (b) reversal of the reflected and refracted beams

The effect of such reversal of the reflected and transmitted beams must be to produce waves of the same amplitudes as those prevailing along the paths of the original beams 1, 2 and 3. Consequently, along the path of beam 1,

$$Ar^2 + Att' = A \tag{7.20}$$

and along path 4,

$$Art + Atr' = 0 \tag{7.21}$$

because there was originally no light along path 4 before reversal. Equation 7.20 gives

$$tt' = 1 - r^2 \tag{7.22}$$

and Equation 7.21 gives

$$r' = -r \tag{7.23}$$

The reflection coefficient of the surfaces is the intensity of the light reflected divided by the intensity of the light incident, where the angles of incidence and reflection are equal for specular reflection. As the intensity is proportional to the square of the amplitude, it follows from Equation 7.23 that the reflection coefficient is the same irrespective of whether the light is incident at the boundary in air or in the medium of refractive index n. In general, the reflection coefficient at the boundary between the two media 1 and 2, of refractive indices n_1 and n_2 respectively, is the same whether the light is incident in medium 1 or in medium 2.

Equation 7.23 shows that the amplitude of the wave in the medium of refractive index n is in the reverse direction to that in air. There is hence a change of phase of π rad; if there is no phase change on reflection at the boundary between media 1 and 2 of refractive indices n_1 and n_2 when the light is incident in medium 1, there will be a phase change of π on reflection when the light is incident in medium 2. Further study making use of electromagnetic theory (Chapter 1, Volume 2) shows, in fact, that the phase change of π occurs when light is incident in medium 1 and $n_1 < n_2$, i.e. when reflection is at the boundary of the medium of higher refractive index and the light is incident in the medium of lower refractive index—corresponding to the light being incident in the medium in which it has the higher speed.

7.8 The Doppler Effect

When a source of waves is in motion relative to a receiver, the frequency as determined by an observer moving with the receiver is different from that determined by an observer moving with the source. If, for example, a whistle is sounded on a train approaching a stationary observer, the frequency of the note will appear to be higher than it actually is (i.e. as determined by an observer on the train), whilst the frequency will appear to be lower after the train has passed the observer and is moving away from him. There is a noticeable fall in the apparent frequency of the sound as the train passes the observer.

This Doppler effect, the difference between the apparent and true frequency brought about by relative motion between the source and observer, is important in wave motion, in particular of sound and of electromagnetic radiation (of which visible light is a part). There is a difference between the Doppler effect in sound and in light however. This is because sound is propagated in a medium (for example, the air) which is considered to be stationary and relative to which the motion of the source or of the observer can be determined. Light waves do not necessitate a medium for propagation, but are readily transmitted through free space. The theory of relativity has, therefore, to be invoked; one of the fundamental tenets of this theory is that the speci- fication of a medium at rest absolutely in which light is propagat- ed is not possible, and only relative motion of the source and observer has significance.

The Doppler effect for waves propagated in a material medium considered to be at rest (as for sound waves in air, assuming there is no wind) is conveniently considered in two stages: the source is in motion with uniform velocity and the observer is at rest in the medium; and the source is at rest and the observ- er is moving with uniform velocity.

7.8.1 THE DOPPLER EFFECT WHEN THE SOURCE IS IN MOTION AND THE OBSERVER IS AT REST

Let S in Fig. 7.10(a) be a point source of waves, and the circle with S as centre denote in the plane of the diagram the section

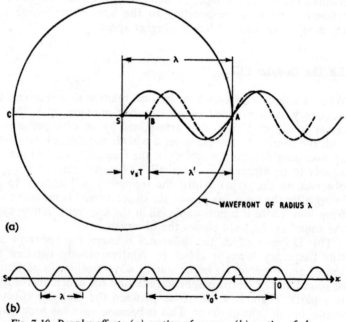

Fig. 7.10. Doppler effect: (a) motion of source; (b) motion of observer

of a spherical wavefront emanating from S. Suppose this wavefront is that prevailing at time T after a wave is given out by the source, where T is the period of the wave motion. If the source S were stationary with respect to the medium, the radius of this sphere would be $vT = \lambda$, where v is the velocity of the waves and λ their wavelength.

Suppose the source S is moving with constant velocity v_s in the medium along the radius SA. In the time T the source will have moved to B, where $SB = v_sT$. During this time the wave motion will still only have spread to A, though the source which produces it is at B at the end of this interval of time. The actual wave will consequently be compressed between BA owing to the motion of the source S. The new wavelength λ', as determined at A and represented by the dotted line in Fig. 7.10(a), will therefore be shorter by the amount v_sT than the wavelength λ [the full line in Fig. 7.10(a)] which would prevail at B if S had been stationary. Therefore,

$$\lambda - \lambda' = v_sT$$

Since $\lambda = vT$, where v is the velocity of the waves (which is decided only by the medium in which they are propagated),

$$\lambda' = (v - v_s)T = \frac{\lambda}{v}(v - v_s)$$

If the source were moving away from A, along radius SC, it is apparent that the new wavelength would be *increased* by v_sT, so would become $\lambda' = \lambda + v_sT$. Therefore,

$$\lambda' = \frac{\lambda}{v}(v \pm v_s)$$

The positive sign holds when the source is moving with velocity v_s directly away from the stationary observer, and the negative sign applies if the source is moving with velocity v_s directly towards the stationary observer.

As already stated, the velocity v of the waves is not affected by the motion of the source as it depends only on the medium (for example, the velocity of sound $v = \sqrt{(\gamma p / \varrho)}$, where p is the pressure of the gas of density ϱ, and γ is the ratio of its specific heat at constant pressure to that at constant volume). Therefore,

$$f' = \frac{v}{\lambda'} = \frac{v}{\lambda(1 \pm v_s/v)} = \frac{f}{1 \pm v_s/v} \qquad (7.24)$$

where f is the frequency of the wave motion emanating from the source as measured by an observer stationary with respect to the source or moving with the source (i.e. f is the true frequency), and f' is the apparent frequency as determined by an observer at rest in the medium but where the source has a velocity v_s.

If v_s is directed towards the observer, the negative sign applies in Equation 7.24, so the apparent frequency f' is greater than the true frequency f; if v_s is directed away from the observer, the positive sign is involved and the apparent frequency is less than the true frequency. When the observer–source direction is not in line with the direction of the source velocity, the change of frequency will be between the two values given by Equation 7.24.

7.8.2 THE DOPPLER EFFECT WHEN THE SOURCE IS AT REST AND THE OBSERVER IS MOVING

Suppose the source of waves S is at rest in the medium but the observer O is moving with a uniform velocity v_o. The source S sends out along the positive direction of the x-axis in Fig. 7.10(b) a train of waves of wavelength λ with velocity v. The observer is at O on this x-axis. If O is stationary also, the frequency of the wave motion observed is

$$f = \frac{v}{\lambda}$$

which is the number of waves which pass O per second.

When the observer O has a uniform velocity v_o along the x-axis *towards* the source, in a time interval t the observer will move a distance $v_o t$ towards S. In doing so, O must encounter $v_o t/\lambda$ additional waves. The number of waves which pass O per second is now f', where

$$f' = \frac{v}{\lambda} + \frac{v_o}{\lambda} = \frac{f}{v}(v+v_o)$$

where f' is the apparent frequency of the source as determined by the moving observer.

If the observer O moves in the opposite direction, away from S, fewer waves are encountered per second than if O were stationary. Now

$$f' = \frac{v}{\lambda} - \frac{v_o}{\lambda} = \frac{f}{v}(v-v_o)$$

Therefore, the equation

$$f' = f\left(1 \pm \frac{v_o}{v}\right) \qquad (7.25)$$

gives the frequency f' of the source as determined by an observer moving with a velocity v_o relative to the stationary source; the positive sign applies when the observer is moving towards the source ($f' > f$), and the negative sign when the observer is moving away from the source ($f' < f$).

7.8.3 THE DOPPLER EFFECT FOR LIGHT

As already mentioned, the only motion which can be specified for light is the relative motion of the source and the observer; the light waves cannot be said to be propagated in a medium regarded as at rest. There is consequently no physical distinction between the two situations discussed in Sections 7.8.1 and 7.8.2. The appropriate equation for the Doppler effect for light and other forms of electromagnetic radiation becomes

$$f' = f \sqrt{\left(\frac{1 + v_r/v}{1 - v_r/v}\right)} \qquad (7.26)$$

where v is the velocity of the light, and v_r is the relative velocity of the observer and the source *towards* one another.

In most terrestrial experiments, the velocity v_r is so small compared with the velocity of light v that the difference between f' and f is negligible. However, a significant difference does occur for light from a gaseous discharge. A high-resolution spectrometer used to observe a single line in the atomic spectrum from a discharge lamp will be subject to broadening due to the Doppler effect. This is because the frequency of the radiation given out by an excited atom in the discharge moving towards the direction of the spectrometer will be greater than that of the frequency due to an atom moving away from the spectrometer; in Equation 7.26, v_r is then the kinetic velocity of the atom, and is positive in the first instance and negative in the second. The wavelength λ' for the atom moving towards the spectrometer (observer) will, therefore, be less than the wavelength λ for an atom stationary with respect to the observer; on the other hand λ'', the wavelength for an atom moving away from the observer, will be greater than λ. The spectrum line of wavelength λ will therefore appear to be spread over the range λ' to λ'' as a result of the Doppler effect.

An aeroplane carrying a radio transmitter will emit, in the forward direction, waves from a source moving at the velocity of the aeroplane. Such waves reflected back from, say, a hillside, will be travelling in the direction opposite to that of the aeroplane. Apparatus within the aeroplane which determines the resulting difference in frequency due to the Doppler effect can be used to measure the aeroplane speed or the distance of the reflector.

The Doppler effect is observed in the light from stars. Stars have velocities towards or away from the Earth between about 10 km s^{-1} and, in some cases, 300 km s^{-1}. Light from a star moving away from the Earth with a relative velocity v_r will be observed on Earth with the frequency decreased and the wavelength increased. If the spectrum of the light from such a star is photographed alongside that of a laboratory source giving the same spectrum lines, the lines in the stellar light will be shifted towards the red end of the spectrum. For a star moving towards the Earth, the shift will be towards the violet end.

In a rather extreme case of a star having a velocity of 300 km s^{-1} away from the Earth, v_r/v in Equation 7.26 is $-300/(3 \times 10^5)$ since the velocity of light in free space is 3×10^5 km s^{-1} approx. Therefore,

$$f' = f \sqrt{\left(\frac{1-10^{-3}}{1+10^{-3}}\right)}$$

$$= f(1-10^{-3}) \text{ approx.}$$

$$= 0 \cdot 999 f$$

The wavelength λ' is, therefore, increased by 1 part in 1,000 approx. compared with the wavelength λ.

Exercise 7

1. Show that the equation $y = a \sin(pt - qx)$ may represent a simple harmonic wave travelling along the x-axis and discuss the significance of p and q.

   Discuss the circumstances in which a group velocity U arises differing from the wave velocity, and show that $U = dp/dq$.

   Sketch a curve to illustrate how the wave velocity for light waves travelling in glass varies with their wavelengths measured in the glass. Show how the group velocity for any wavelength may be obtained from the tangent to this curve at the appropriate point. (L. P.)

2. Explain the meanings of the terms *phase* (or *wave*) *velocity* and *group velocity* in relation to wave propagation in a dispersive medium. Show that the phase velocity $V = v/k$, whilst the group velocity $U = dv/dk$,

where ν is the frequency and $k = 1/\lambda$, λ being the wavelength in the medium.

Find the relationship between the group velocities of light *in vacuo* and in a dispersive medium of which the refractive index μ is given by

$$\mu = A + B\nu^2$$

If, for carbon disulphide, $A = 1.507$ and $B = 0.222 \times 10^{-30}$ sec², find the group velocity for light that in a vacuum has a wavelength of 5893 Å. (Take the velocity of light *in vacuo* $c = 3.000 \times 10^{10}$ cm sec⁻¹.) (L. G.)

3. By considering the superposition of two progressive sine waves of equal amplitude but slightly different frequency, explain the distinction between phase and group velocity and derive an expression for their difference. Calculate to three significant figures the velocity of light of wavelength 5000 Å in carbon disulphide at 20°C for which the refractive index μ is given by $\mu_{20} = 1.577 + (1.78 \times 10^{-10})/\lambda^2$, where λ is in cm. Assume that the velocity of light *in vacuo* is 3×10^{10} cm sec⁻¹. (L. P.)

4. Explain how the Doppler effect influences the radiation received from terrestrial and astronomical sources.
What steps can be taken to minimise its effect in terrestrial sources and to what useful information does its presence lead in astro-physics? (L. P.)

5. How does the Doppler effect influence the positions and widths of spectral lines?
Briefly review what information can be derived from the Doppler effect measurements in astronomy and astro-physics. (L. P.)

6. A line at a wavelength of 6,000 Å in the spectrum of the light from a star is observed to be 0.1 Å longer in wavelength than the light in the corresponding spectrum line from a terrestrial source. Assuming that this increase is due to the Doppler effect, calculate the velocity with which the star is receding from the Earth. (The velocity of light in free space $= 3 \times 10^{10}$ cm s⁻¹.)

CHAPTER 8

Interference between Two Beams of Light Involving Division of Wavefront

8.1 Coherent Disturbances: Division of Wavefront and Division of Amplitude

Experiments on the interference of light involve the superposition of wave motions (Section 7.3). The superposition of two simple sinusoidal waves of the same frequency, where one has an amplitude A_1 and the other A_2, results in an intensity given by Equation 7.15 as

$$I = A_1^2 + A_2^2 + 2A_1A_2 \cos \delta$$

where I, the intensity, is equal to the square of the resultant amplitude and δ is the phase difference between the two waves in the region where they are superposed.

The disturbances emitted by a source of light undergo rapid random changes of phase. For example, substantially monochromatic radiation can be obtained from a discharge lamp source, such as a sodium lamp or a cadmium lamp. This radiation is due to energy changes resulting from electron transitions within the atoms of the excited element. In the discharge, the atoms will radiate energy independently of one another without any definite phase relationships between the successive wave trains of finite length that are emitted. The abrupt random changes of phase will occur at the rate of 10^8 per second or more. If, therefore, two independent sources (for example, two cadmium lamps) are used in an attempt to observe interference phenomena, the phase difference δ in the above equation will change millions of times per second. Neither the eye nor any faster acting device, such as a photoelectric cell, can follow such rapid intensity variations; consequently, the mean value of cos δ (which is equal to zero) will prevail in the observation, and the mean intensity will be $A_1^2 + A_2^2$ with no perceptible interference.

To obtain the necessary constant difference of phase between the disturbances in the interfering beams, it is essental to use a single source and to divide the light emitted from it into two parts. The phase changes due to the source in the one beam will then be simultaneous with those in the other. If the two beams are superposed after traversing slightly different optical path lengths, the corresponding time difference will be so short because of the enormous speed of light that simultaneity in phase changes is still a valid assumption. Two or more disturbances which have a constant phase relation between them are coherent; coherent disturbances will interfere. When no such constant phase relation exists, the disturbances are incoherent and interference observations are not possible.

In interference methods involving division of wavefront, light from a single source is divided into two beams by the use of apertures or optical components which separate neighbouring parts of the wavefront; the separated beams are superposed subsequently. A point or line source of light is involved. Examples of such experiments on interference are those using Young's two slits, Fresnel's biprism, Fresnel's double mirror and Lloyd's single mirror. In division of amplitude, a single beam is divided by means of partial reflection in a thin plate, film or semi-transparent mirror, and the divided beams are subsequently brought together. Division of amplitude methods with two beams are described in Chapters 9 and 10.

It is not strictly satisfactory to isolate interference phenomena from diffraction. When an aperture intercepts light waves, light appears within the geometrically defined shadow: diffraction occurs. If the aperture is large—as, for example, when a semi-transparent reflector is used in amplitude division—the effect of diffraction is minor and can be ignored.

If the aperture is small, diffraction is important. For example, in Young's experiment, two slits are employed to give division of wavefront. Diffraction effects due to the small finite widths of these slits will occur, as well as interference. In a first study, it is a convenient simplification to disregard the diffraction and look upon the slits as two line sources of light which produce coherent disturbances because they are illuminated by a single source.

It is arguable as to whether interference or diffraction should be considered first: logically, diffraction is inevitably present and should thus receive prior treatment; from the point of view of simplicity, it is preferable first to deal with interference.

8.2 Young's Experiment

In 1807, Thomas Young performed one of the first experiments on interference: sunlight was arranged to pass through a pinhole S to produce spherical wavefronts which impinged upon two separate pinholes A and B in a screen some distance away [Fig. 8.1(a)]. The two sets of spherical waves from A and B interfered to produce a symmetrical pattern on the viewing screen Z. In

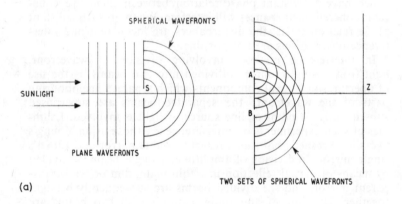

(a)

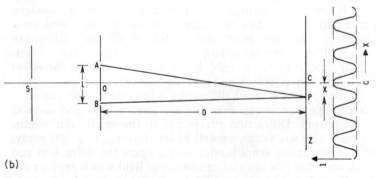

(b)

*Fig. 8.1. Young's experiment on interference: (a) the original arrangement;
(b) the more recent double-slit method*

later versions of this experiment, the pinhole is replaced by a slit
S [Fig. 8.1(b)] which is illuminated by a monochromatic source—
conveniently a sodium lamp—and the pinholes A and B are replaced by parallel slits equidistant from S; the slits are frequently
made by cutting thin lines about 1 mm apart into an opaque screen
in the form of a uniformly exposed and developed photographic
plate.

If the slits A and B are of the same width and equidistant from
the illuminated slit S, they will act as coherent sources of substantially monochromatic light in the same phase and of equal
amplitudes. At the screen Z (ideally the graticule of a micrometer
eyepiece), there will appear a set of straight interference fringes,
alternately light and dark and parallel to S, A and B.

Let the separation between the slits A and B be l, the distance
$OC = D$ (where O is midway between A and B, and C is the point
on the screen Z equidistant from A and B, so OC is perpendicular
to AB and to the plane of screen Z). The appearance at any point
P on the screen which is separated from C by a distance x will
be decided by the superposition at P of light disturbances emanating from A and B. If x is small compared with D, the distances
AP and BP are so nearly equal and nearly parallel that the amplitude of the wave arriving at P from A may be assumed to be
equal to that from B and also to act in the same direction. At
P, therefore, these waves will interfere. How they interfere will
depend on the difference between the optical paths AP and BP
which, though small, is significant compared with the wavelength
λ of the monochromatic light employed to illuminate slit S.
Clearly,
$$AP^2 - BP^2 = [(0{\cdot}5l+x)^2 + D^2] - [(0{\cdot}5l-x)^2 + D^2]$$
$$= 0{\cdot}25l^2 + lx + x^2 + D^2 - 0{\cdot}25l^2 + lx - x^2 - D^2$$
$$= 2lx$$

Therefore,
$$AP - BP = \frac{2lx}{AP+BP} = \frac{2lx}{2D}$$

because $AP + BP \simeq D$.

The waves from the slits will interfere constructively to give
bright bands in the interference pattern on the screen Z when the
path difference $AP - BP = n\lambda$ or zero, where n is an integer. The
zero will occur at the centre point C, which will thus be bright.
Destructive interference resulting in dark bands will occur when
this path difference is $(n+0{\cdot}5)\lambda$. Hence,
$$\frac{lx}{D} = n\lambda$$

or

$$x = \frac{nD\lambda}{l} \tag{8.1}$$

gives the positions x of the maxima in the bright bands in the interference pattern, whereas

$$x = \frac{(n+0\cdot5)D\lambda}{l}$$

gives the positions of the zero intensities in the dark bands.

The phase difference δ between the two light disturbances arriving at P due to the path difference of lx/D is clearly $2\pi lx/\lambda D$. The intensity I at P is, therefore, given by Equation 7.18 as

$$I = 4A^2 \cos^2\left(\frac{\pi lx}{\lambda D}\right) \tag{8.2}$$

where A is the amplitude of either beam. The intensity range in the interference pattern is consequently from a maximum of $4A^2$, corresponding to an amplitude of $2A$, to a minimum of zero. The variation of intensity following this cosine-squared distribution is indicated on the right-hand side of Fig. 8.1(b).

In a students' experiment, Young's slits were used to find the wavelength λ of light from a sodium lamp (nominally 5,893 Å). The distance D was 30 cm, l was 1·0 mm and λ was found to be $(5,865 \pm 60)$ Å.

If white light is used instead of monochromatic light, the central bands (three in number with $l = 1$ mm and $D = 200$ cm) will be white and the others tinged with colour until they merge further from the centre into a uniform grey. The inside edges of the bands are coloured blue, and the outer edges are coloured red. After a few bands, the blue of band $n+1$ overlaps with the red of band n.

Young's experiment was a vital one in establishing the wave theory of light as opposed to the earlier corpuscular theory of Newton. It received criticism, however, on the basis that the edges of the slits used might modify the light in a manner which led to the observed pattern being produced by other than interference between light waves. Within a few years, Young's explanation was justified by the experiments of Fresnel with his biprism and double mirror. These also have enabled a much brighter interference pattern to be obtained (this was faint in the experiment of Young), and the biprism experiment is a particularly valuable one for students.

8.3 Fresnel's Biprism

The isosceles biprism used has a refracting edge F (Fig. 8.2), at which the angle between the two faces EF and FG is about $1°$ less than $180°$. The face EG of this prism is plane. The edges E and G are blunt to avoid easily damaged sharp ends. A vertical slit S is mounted on an optical bench parallel to the vertically mounted refracting edge F of the biprism. The slit S is illuminated, usually by monochromatic sodium light in a students' experiment. Beams from S impinging on faces EF and GF are deviated by the biprism and so appear to come from lines A and B after refraction by the biprism. Thus A and B act as coherent line sources producing an interference pattern at the distant screen Z. The separation between A and B has to be measured as described later; this is best done first in the experiment. Observation and measurement of the interference fringes is usually by means of a low-power microscope with a micrometer scale.

The pattern of interference fringes occurs only over the region YY of the screen Z where the waves overlap. The interference pattern obtained shows sharp equally spaced parallel fringes, superimposed upon which is a more diffuse and gradual variation in intensity. This gradual variation is due to diffraction of the light by the straight edge which is formed by the vertex F of the biprism.

The diffraction results in the broader structure of the pattern but is ignored in considering the well-defined interference fringes proper.

When setting up the biprism on an optical bench, it is essential to ensure that the refracting edge F is accurately parallel to the illuminated slit S. If it is not, a neighbouring point on S—inevitably of finite width—will also produce interference fringes, the two fringe systems will have maxima which do not coincide,

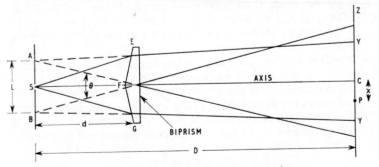

Fig. 8.2. Fresnel's biprism experiment on interference

resulting in an interference pattern lacking in sharpness and contrast.

If D is the axial distance between the slit S and the screen Z (i.e. if $SC = D$), the virtual images of the slit at A and B are separated by a distance l, and λ is the wavelength of the monochromatic light used, Equation 8.1 applies to give

$$x = \frac{nD\lambda}{l}$$

where x is the separation between a bright fringe at a point P on the screen Z and the centre point C, and n is an integer. The separation s between neighbouring centres of bright fringes in the interference pattern, therefore, corresponds to a change of n to $n+1$; so

$$s = \frac{D\lambda}{l} \tag{8.3}$$

With a microscope furnished with a micrometer eyepiece, a number of values of x is found by measuring the separation between a counted number of bright (or dark) fringe centres, and s is then determined by simple division. A number of observations are necessary, as it is likely to be difficult to judge where the cross-wires are on the centre of a fringe.

It is necessary to measure l; there are two ways of doing this.

1. Place a converging lens between the biprism and the micrometer eyepiece. Move the lens laterally along the optical bench so that sharp images of the slit are formed in the eyepiece. There are two positions of the lens at which such images are obtained. As the minimum separation between the object and real image is four times the focal length of the lens, it is essential to choose a lens of suitable focal length and also to carry out this measurement *before* the interference study is undertaken (otherwise it may be found that the separation D used is too small; also, the lens has to be situated in its two positions between the biprism and the screen Z).

   The separations between the images of the slit are measured by the microscope micrometer at each of the two focused positions. If u is the object distance (from the slit S to the lens) in the first position of the lens and v the image distance (from the lens to the focal plane of the microscope), the separation C_1 between the images of the slit will be given by

$$C_1 = \frac{lv}{u}$$

In the conjugate second position of the lens, u becomes v and vice versa; the separation C_2 of the slit images determined by the microscope is now

$$C_2 = \frac{lu}{v}$$

Therefore,

$$C_1 C_2 = l^2$$

Thus, $l = \sqrt{(C_1 C_2)}$ and, as C_1 and C_2 are measured, l is found.

2. Considering the biprism as two thin prisms each of refracting angle α and of glass of refractive index n, it is readily shown that

$$l = d\theta = 2d(n-1)\alpha$$

where $d = SF$, measured directly on the scale of the optical bench, whilst n and α are measured in a separate experiment with a spectrometer.

8.4 Fresnel's Double-mirror System

Two plane mirrors M_1 and M_2 (two pieces of good quality plate glass each about 5 cm square are satisfactory) are placed together, with their planes at a small angle θ to one another and their contiguous edges forming a straight line of intersection through O, as shown in Fig. 8.3. A slit S illuminated with monochromatic

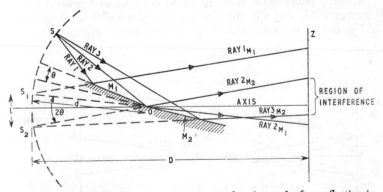

Fig. 8.3. Fresnel's double-mirror system: ray 1_{M_1} is ray 1 after reflection in M_1; ray 2_{M_1} is ray 2 after reflection in M_1: ray 2_{M_2} is ray 2 after reflection in M_2; ray 3_{M_2} is ray 3 after reflection in M_2

light of wavelength λ is set up accurately parallel to the line of intersection of the mirrors, so that light reflected from the mirrors is at near grazing angles (these angles are exaggerated for clarity in Fig. 8.3). The arrangement can then be conveniently set up on an optical bench. Light from the single slit S is separated into two beams by reflection from the two mirrors and, on the section of the observing screen Z where these beams overlap, interference fringes are seen. The light rays interfering at the screen appear to come from virtual images S_1 and S_2 formed by reflection of S in the mirrors M_1 and M_2 respectively. The formation of the interference pattern is explained as for the Young's double-slit experiment and the biprism. As before, if l is the separation between the virtual sources S_1 and S_2, the axial distance from the plane of S_1S_2 and the screen Z is D, and λ is the wavelength of the monochromatic light used, Equation 8.3 applies to give the separation s between neighbouring centres of light (or dark) fringes. Further, l can be found because $l = 2d\theta$, where $S_1O = d$ and the angle subtended by S_1 and S_2 at O it twice the angle between the mirrors (which is determined in a separate experiment with a spectrometer).

8.5. Lloyd's Single-mirror Experiment on Interference

Rays from an illuminated slit S fall on the plane mirror MN at grazing incidence and are reflected to the screen Z or the eyepiece, where they interfere with the direct rays from S (Fig. 8.4). The coherent sources of light giving rise to interference are thus the slit S and its mirror image S'. A suitable mirror is a piece of selected plane plate glass about 30 cm long and several centimetres wide. The slit must be accurately parallel to the plane of the mirror, to ensure that it is also parallel to its image.

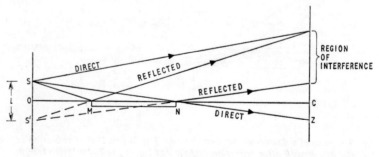

Fig. 8.4. Lloyd's single-mirror experiment

If the separation between the slit and its virtual image in the plane mirror is l, and the axial distance from the plane of SS' to the plane Z is $OC = D$, Equation 8.3 gives the separation s between the centre of neighbouring light or dark fringes, where λ is the wavelength of the monochromatic light used.

Lloyd's mirror experiment dates from 1834. The fringes are easily found. Owing to reflection, the sources are laterally reversed, so the lower part of slit S (necessarily of finite width) will emit waves in the same phase as the upper part of the slit image S'. Consequently, Lloyd's mirror gives sharper fringes than Fresnel's mirrors for a considerable width of slit. This aids their observation: wider slits enable more light to be produced in the interfering beams.

As there is no reflection from the underside of the mirror, only half the interference pattern is seen as compared with the previously described arrangements. To observe the 'central' fringe—central if the whole pattern were observable and preferably called the fringe for zero path difference—it is necessary to move the observing screen Z (or eyepiece) to be nearly in contact with the end N of the mirror. Lloyd found that this fringe of zero path difference was dark instead of bright. This confirms experimentally that there is a phase change of π when light is reflected at a denser medium (Section 7.7). Thus, a path difference of zero or $n\lambda$ is the condition for a dark fringe, where n is an integer, whereas a bright fringe is obtained when the path difference is $(n+0.5)\lambda$.

Exercise 8

1. What is meant by interference in wave motion?
   Describe how the wavelength of the mercury green line (5461 Å) may be measured using a Fresnel biprism.
   During such an experiment a thin transparent film placed in the path of the light from one of the prisms causes a shift of 5·0 fringes. If the film has a refractive index of 1·33, calculate its thickness.   (L. G.)

2. Describe and give the theory of a method of determining the wavelength of the green light from a mercury lamp.   (L. Anc.)

3. Give a simple theory of the formation of the interference fringes produced by Fresnel's biprism. Indicate the nature and origin of the difference between the ideal theoretical interference pattern and the actual fringe system observed. Explain how the wavelength of visible light from a monochromatic source may be determined by measurements of interference fringes from a Fresnel biprism. Discuss carefully the principal sources of error in this experiment.   (L. G.)

4. A radio-telescope placed on a cliff top forms with the surface of the sea a 'Lloyd's mirror' to waves from a point-like radio star near the horizon. Obtain an expression for the intensity of the signal received in terms of α, the angle characterising the height of the radio star above the horizon, h, the height of the radio-telescope above the surface of the sea and λ, the wavelength to which the receiver is tuned. (Assume perfect reflection from the surface of the sea.)

If two such sources of equal intensity close to the horizon ($\sin \alpha = \alpha$) and lying in the same vertical plane are separated by $\Delta\alpha$ show that there will be no variation in the signal strength with α if $\Delta\alpha = \lambda/4h$.   (L. P.)

CHAPTER 9

Interference between Two Beams of Light: Division of Amplitude by Parallel Plate or Film

9.1 Interference of Light in Plane Parallel-sided Plates

Interference involving division of amplitude is demonstrated by the use of a parallel-sided transparent plate of moderate thickness. It must be recalled that, at this stage, the interference of only two beams is being considered. If the plane-parallel sides of the plate have significant reflection coefficients, (for example, because they are silvered or aluminised), several beams will be involved in determining the interference pattern. This subject of multiple-beam interference is deferred until Chapter 4, Volume 2.

To establish simply an important basic equation, consider a parallel-sided plane plate of transparent material of refractive index n and of thickness e. Let a ray of monochromatic light of wavelength λ from the point source P be incident on the surface of the plate in air at an angle I at point A (Fig. 9.1). If I is considerably less than $90°$, a small fraction of this light is reflected along AB, and the remainder is refracted into the material along AC, where the angle of refraction is R; at C, part of the light is reflected along CD to emerge at D along DE, and part is transmitted along CF. Further successive reflections and refractions will occur to the right of C and D in Fig. 9.1 so, strictly, multiple

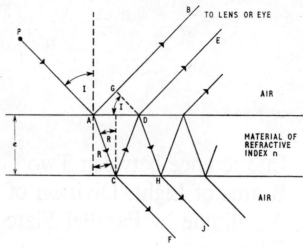

Fig. 9.1. Single ray of light incident upon a plane parallel-sided transparent plate

beams will emerge by reflection and by transmission. These subsequent beams are weak (with rapidly decreasing intensities) so may be ignored at first; however, the reflected beams AB and DE have similar intensities as each has undergone one reflection and the light absorption in the material is presumed to be small.

There will be an optical path difference between the emergent reflected rays AB and DE. To find this difference, DG is constructed perpendicular to AB. DG is then a wavefront in the reflected parallel light formed by rays AB and DE. Light in ray AB has to travel the distance AG from A to meet this wavefront. Meanwhile, light in ray DE has had to traverse the optical path $AC + CD$ in the transparent material to reach D in the same wavefront. The optical path difference is, therefore,

$$(AC + CD)n - AG$$

From the geometry of Fig. 9.1, it is readily seen that

$$AC = CD = e \sec R = \frac{e}{\cos R}$$

and

$$AG = AD \sin I = 2e \tan R \sin I$$

From Snell's law of refraction,

$$\sin I = n \sin R$$

Therefore,

$$AG = 2ne \tan R \sin R = \frac{2ne \sin^2 R}{\cos R}$$

The optical path difference is, therefore,

$$\frac{2ne}{\cos R} - \frac{2ne \sin^2 R}{\cos R} = \frac{2ne}{\cos R}(1 - \sin^2 R) = 2ne \cos R \quad (9.1)$$

There is a phase change of π on reflection at a denser medium (Sections 7.7 and 8.5). The reflection at A involves this phase change: the reflected ray AB is subject to it, but the ray DE is not. It follows that the disturbances in the wavefront GD and the rays AB and DE are in phase if

$$2ne \cos R = (p + 0\cdot 5)\lambda \quad (9.2)$$

and in anti-phase if

$$2ne \cos R = p\lambda \quad (9.3)$$

where p is an integer. With perpendicular incidence, $I = R = 0$, and $\cos R = 1$, so Equation 9.3 becomes

$$2ne = p\lambda \quad (9.4)$$

So far, only a single incident ray from the point source P has been considered. In fact there will be a cone of diverging rays from P incident at various angles I. If a lens of small aperture (of which the human eye is an example) is placed to receive the light along a particular direction (for example, the beam of rays between AB and DE in Fig. 9.1), the range of values of angles of I and R is restricted; consequently, in the focal plane of the lens or at the retina of the relaxed normal eye, the intensity will be a maximum if Equation 9.2 holds and a minimum for Equation 9.3.

As the aperture of the lens used to collect the reflected light is increased, rays will enter the lens from different parts of the parallel-sided plate. Particular pairs of parallel rays entering the lens will give maxima or minima for particular values of R in accordance with Equations 9.2 and 9.3 respectively. As there is now a range of possible values of R, so there will be a variation of intensity across the focal plane of the large-aperture lens.

In the important case when P is replaced by an extended source of light (for example, a diffusing screen illuminated by a sodium lamp), a large number of separate point sources is in effect concerned (Fig. 9.2). Suppose the relaxed or unaccommodated eye is viewing the light reflected from the plate from a particular

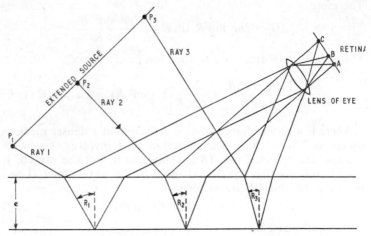

Fig. 9.2. Light from an extended source incident upon a plane parallel-sided transparent plate

position. Consider those points P_1, P_2 and P_3 on the extended source. Ray 1 from P_1 will give rise to a maximum or minimum at point A on the retina, depending on whether $2ne \cos R_1 = (p+0.5)\lambda$ or $p\lambda$ respectively; likewise, for point B on the retina with respect to point P_2 on the extended source and angle R_2, and for C on the retina relative to P_3 and angle R_3. There is consequently a variation in brightness of field across the retina, depending on the direction of observation, there being a continuous varation of the angle R from that corresponding to P_1 at one extreme end of the extended source to that corresponding to P_3 at the other extreme end.

Fig. 9.2 represents only a two-dimensional view of the practical three-dimensional situation. In the field at the retina of the eye (or in the focal plane of a converging lens) when the monochromatic source is extended, at any particular point there will be a maximum brightness where

$$2ne \cos R = (p+0.5)\lambda$$

Both n and e are constant for a given plate, λ is constant for a given monochromatic source of light, and R will vary depending on the direction of observation. A particular value of R will be maintained for a circle in the focal plane of the observing lens or at the retina of the eye. Hence, bright interference *fringes of equal inclination* (R = constant) will be observed in the form of a series of concentric circles (or arcs of circles, depending on the viewing conditions) interspaced by dark fringes where

$$2ne \cos R = p\lambda$$

The fringes can be viewed by transmission as well as by reflection. For transmitted rays (for example, along CF parallel to HJ in Fig. 9.1), no phase change of $90°$ on reflection at an optically denser medium is involved; therefore the appropriate equations will be, for a maximum,

$$2ne \cos R = p\lambda \qquad (9.5)$$

and, for a minimum,

$$2ne \cos R = (p+0.5)\lambda \qquad (9.6)$$

These equations are complementary to those for reflected rays.

The fringes of equal inclination observed by transmission exhibit poor contrast compared with the reflected ones. This is because neighbouring transmitted rays CF and HJ in Fig. 9.1 are unequal in amplitude: CF has not undergone reflection, whereas HJ has suffered two internal reflections at each of which the reflection coefficient is only about 0.1 unless the plate surfaces are silvered or aluminised—and then multiple-beam interference has to be considered. Thus, the transmission field pattern shows variations in intensity obscured by superimposed uniform light. On the other hand, neighbouring reflected rays AB and DE have each undergone one reflection and are of similar, although not the same, amplitude.

In the particular case when near-normal incident light and an extended source (or a point source object relative to an auxiliary converging lens) are used, the concentric circular alternately light and dark fringes in the interference pattern are called *Haidinger's fringes*. They can be observed either by reflection [when a reflector plate at $45°$ is needed so that the observer does not obstruct the field: cf. Fig. 9.4(a)] or by transmission.

For fringes of equal inclination observed in a parallel-sided plate, if e is large (i.e. if the plate is thick—of the order of 0.5 cm or more) it is seen from Equation 9.2 that the fringes are very close together. Thus, two neighbouring bright circular fringes corresponding to a change from p to $p+1$ are separated by angles of refraction (and so angles of incidence and reflection) corresponding to a change of $\cos R$ from $(p+0.5)\lambda/2ne$ to $(p+1.5)\lambda/2ne$ (i.e. $\lambda/2ne$) which is a small change of R if e is large. To enable such close fringes to be observed, satisfactory magnification by a suitable telescope of sufficiently wide aperture to accommodate neighbouring rays separated by GD (Fig. 9.1) is demanded.

On the other hand, if e is very small (i.e. if the plate is a *thin film* of, for example, mica, plastic, air between plates, a soap solution film, or oil on water), AD and GD in Fig. 9.1 are small distances, so rays AB and DE follow such closely neighbouring paths that they interfere to give one ray emerging from the point A. Fringes are then readily observed by the unaided eye and, furthermore, exhibit the effects of local variations in film thickness more noticeably than variations of the angle of emergence of the reflected (or transmitted) light. Now a considerable change of $R = \cos^{-1}(\lambda/2ne)$ in moving from one fringe to its neighbour is needed when e is small. Indeed, if e is exceedingly small, variations of R become so slight as to give an appearance of uniform intensity with a moderately extended source unless there are also significant variations of film thickness.

9.2 Interference in Thin Films: The Wedge Film

The concepts of Section 9.1 can be applied to explain the colouration of very thin films when illuminated by white light. The thickness of the film will, in general, vary and the optical path difference in the rays observed by the eye, which are the externally and internally reflected rays, will be decided by the thickness of the film in the immediate locality of the point at which reflection occurs—i.e. point A, almost coincident with point D, Fig. 9.1. Effects of variation of the angle I will be much less significant than thickness changes arising away from A. Dark bands appear when $2ne \cos R = p\lambda$ (Equation 9.3); now variation of the thickness e is important but not that of angle R. With white light, destructive interference between neighbouring reflected beams from a point of emergence occurs at particular values of e, largely independent of the angle of emergence I where $\sin I = n \sin R$. At such points, the field of view is coloured with the complementary colour formed by the remainder of the spectrum (Table 9.1). The contour of abrupt thickness changes is thus delineated.

Table 9.1. COMPLEMENTARY COLOURS OF THE SPECTRUM

Colour	Red	Orange	Yellow	Green	Blue	Violet
Complementary colour	Blue-green	Blue	Indigo	Purple	Orange	Green-yellow

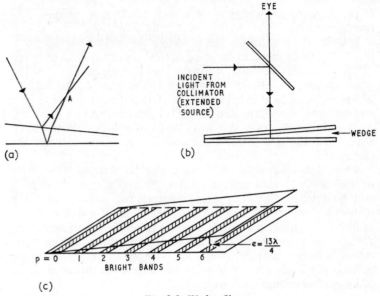

Fig. 9.3. Wedge films

Variations of thickness over the film surface are sometimes gradual: a section of the film will then be wedge-shaped. An example is a film of oil on water. Colouration appears as before but the rays reaching the eye appear to diverge from point A [Fig. 9.3(a)], which is only slightly above the surface of the wedge because e is very small.

A wedge film of air between plane sheets of glass (microscope slides will serve) is conveniently viewed with normal incident light by the arrangement shown in Fig. 9.3(b), where an auxiliary thin glass reflecting plate set at 45° to the direction of viewing enables the incident light to be conveniently from a source to one side. With monochromatic light of wavelength λ, dark bands appear when $p\lambda = 2e$, bright bands when $(p+0\cdot5)\lambda = 2e$, and the first bright band occurs near the apex of the wedge at $p=0$, i.e. when $e=\lambda/4$ [Fig. 9.3(c)]. With normally incident white light, again the complementary colour occurs wherever a particular wavelength λ is subject to destructive interference. In Fig. 9.3(c), with white light, the thin edge is dark (as it is also with monochromatic light); then, moving towards the right-hand side, a violet fringe appears followed by other colours in the order of the spectrum. After a short distance, depending on the angle of the wedge, the field of view assumes a uniform grey appearance as bright fringes of one wavelength coincide with dark fringes of others.

If the angle of the wedge is decreased, the fringes become further apart.

The use of a true optical flat in testing the surface contours of a piece of glass or metal undergoing optical polishing is an example of the principles outlined. With the true test flat placed over the work surface, an interference pattern is observed. As the flatness of the work surface is improved, the fringes move further apart and become of a shape depending on the contours of the work surface. Prominences in this work surface are then polished down until, with skilled technique, the interference fringe contours are removed.

A small distance d, in the form of the diameter of a thin wire or thickness of a thin strip of solid material (for example, paper), may be determined if it is placed between two rectangular plane plates of glass to maintain the right-hand end of a wedge of air; as in Fig. 9.3(b). It is a simple exercise to relate d to its measured distance from the thin end of the wedge and the measured separation between neighbouring dark fringes in the interference pattern obtained when normally incident monochromatic light of known wavelength λ is used.

9.3 Newton's Rings

Interference occurs in an air or liquid film between the convex curved surface of a lens and a flat glass plate on which it rests. In the pioneer observation of the concentric circular interference fringes by Newton, he used lenses with radii of curvature of 28 ft and 51 ft so that the rings would be large enough to be measured without magnification. With a lens of radius of curvature 410 cm, which is large in the usual laboratory stock, and with red light, the diameter of the first few bright rings are 2·4 mm, 5·2 mm and 6·4 mm. The usual experimental study of Newton's rings therefore involves a microscope to measure the ring diameters. The arrangement is as in Fig. 9.4(a), with a thin plate glass reflector at 45°; this reflector enables the collimated monochromatic light to be incident from a source to one side, and the microscope to receive light normally from the horizontal flat glass plate beneath the plano-convex lens. The Newton's rings appear as alternately bright and dark circles around the point of contact of the lens and the plate. As there is a phase change of 180° in the light reflected from the surface of the flat plate, the central spot is dark.

From Equation 9.3, for $R = 0$ so $\cos R = 1$, the condition for destructive interference between light rays reflected from the

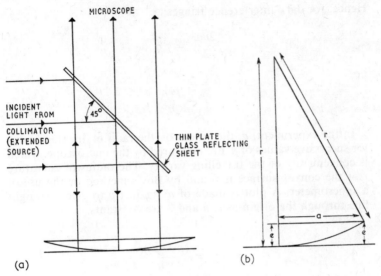

(a)

(b)

Fig. 9.4. Newton's rings

curved surface of the lens (forming the top surface of the film of air or liquid between the lens and plate) and the surface of the flat plate (forming the bottom surface of the film) is

$$2ne = p\lambda$$

where n is the refractive index of the material of the film of thickness e and p is an integer; the condition for constructive interference is:

$$2ne = (p+0\cdot5)\lambda$$

In this arrangement, e is not constant but increases with the radius a between the central point of contact of lens and plate and the point in the field of view concerned [Fig. 9.4(b)]. If r is the radius of curvature of the convex surface of the plano-convex lens, it is readily seen from Fig. 9.4(b) that:

$$r^2 = (r-e)^2+a^2 = r^2-2re+e^2+a^2$$

As e^2 is negligibly small, therefore,

$$e = \frac{a^2}{2r}$$

Hence, for dark interference fringes,

$$\frac{2na^2}{2r} = p\lambda$$

i.e.

$$a^2 = \frac{pr\lambda}{n} \qquad (9.7)$$

In the experiment, a, the values of the radii of the dark rings for successive values of p, are measured by the microscope, which is conveniently of the travelling variety. The radius of curvature r of the convex surface is found by Boy's method or the use of a spherometer. A plot is made of a^2 against p to give a straight line through the origin as r, n and λ are constants.

Now,

$$a_1^2 - a_2^2 = \frac{(p_1 - p_2)r\lambda}{n}$$

where a_1 and a_2 are values of a for fringe numbers p_1 and p_2, so the gradient m of the straight line graph is obviously given by the equation

$$m = \frac{a_1^2 - a_2^2}{p_1 - p_2}$$

whilst $\lambda = mn/r$. With $n = 1$ for air, λ (the wavelength of the light used) can be found.

If the film between the lens and plate is first air (for which the slope m found experimentally is m_a) and second a liquid of refractive index n relative to air as unity (for which $m = m_l$),

$$m_a = \lambda r$$

since $n = 1$, and

$$m_l = \frac{\lambda r}{n}$$

Therefore,

$$n = \frac{m_a}{m_l}$$

Hence, the refractive index of the liquid can be found by determining the slope of the graphs plotted for the film in air and then the film in liquid.

9.4 Anti-reflection and High-reflection Films: Blooming

When monochromatic light of wavelength λ is normally incident in air upon the surface of glass of refractive index n at the wavelength λ, the ratio of the reflected and incident amplitudes is $(n-1)/(n+1)$ (see Chapter 1, Volume 2). For glass of refractive index 1·5, this ratio is $0·5/2·5 = 1/5$, so the ratio of the intensities is $1/25$ (i.e. 4% of the incident light is reflected). This amount will increase with increasing angle of incidence. Consequently, some light is lost by back-reflection in passing through a lens. In a compound lens system, such as that of a camera or microscope objective, containing six or seven components, the additive losses at successive reflecting surfaces may amount to 25% or so. This is not serious as regards light transmission, but the back-reflected light may well cause undesirable effects if it reaches the image formed by the lens. In a camera, for example, it will result in some fogging of the photographic plate or film and there will be a lack of contrast in the image; moreover, glare spots may occur.

This back-reflected light can be substantially reduced by coating the glass surface with a uniform film of transparent material of optical thickness $\lambda/4$ and refractive index $\sqrt{n}$, where n is the refractive index relative to air of the glass so coated. The technique of depositing such anti-reflection films on the glass by evaporation in a vacuum is known colloquially as 'blooming'.

That the back-reflected light is eliminated is seen from Fig. 9.5, which deals with the simple ideal case where the monochromatic light of wavelength λ is incident normally on a plane glass plate. The refractive index of the film material is n_1, whilst that of the glass substrate is n, both relative to air. Let the incident beam be beam 1, the beam reflected from the top surface of the film be beam 2, and beam 3 that reflected from the interface between the film and glass substrate. (Beams 1, 2 and 3 are separated for clarity in Fig. 9.5.) There is a phase change of π between beams 2 and 1 because of reflection at an optically denser medium; there is also a phase change of π between beams 3 and 1 if $n_1 < n$. The optical path difference between reflected beams 2 and 3 results from the fact that beam 3 traverses the film twice. If the optical thickness of this film is $\lambda/4$, this corresponds to a phase change of π. Consequently, reflected beams 2 and 3 follow the same path and are in anti-phase with one another, so will interfere destructively. If they also have the same amplitude, they will annul one another and the back-reflection will be zero. Let the amplitude of beam 1 be A. The amplitude of reflected beam 2 is therefore $A(n_1-1)/(n_1+1)$, whilst that of beam 3 is

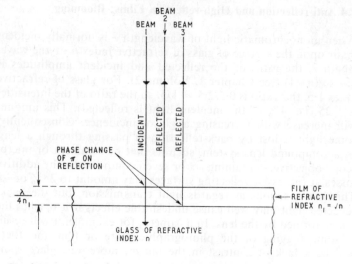

Fig. 9.5. Anti-reflection film on a plane glass surface (light of wavelength λ is incident normally)

$A(n-n_1)/(n+n_1)$, assuming no light absorption in the film. For the amplitudes of beams 2 and 3 to be equal, it follows that:

$$\frac{n_1-1}{n_1+1} = \frac{n-n_1}{n+n_1}$$

Therefore,

$$(n_1-1)(n+n_1) = (n_1+1)(n-n_1)$$

i.e.

$$n_1^2 = n$$

or

$$n_1 = \sqrt{n}$$

For glass of refractive index 1·5, the anti-reflection film should, therefore, have an optical thickness of $\lambda/4$ and a refractive index of $\sqrt{1·5} = 1·225$. Suitable vacuum-deposited films are those of magnesium fluoride (which, after baking in air at 300°C, has a refractive index n_1 of 1·37) and of cryolite, which has a refractive index of 1·34. The latter is thus more suitable optically but it is not as durable mechanically as magnesium fluoride. Clearly, therefore, anti-reflection films are not entirely satisfactory with glasses except those of the higher refractive indices. However, anti-reflection coatings are often prepared by vacuum-deposited magnesium fluoride, although more recent methods give more effective anti-reflection by the use of two films, one deposited on top of the other on the glass substrate.

Films on glass can also be prepared to enhance the reflection. This is done by the deposition of a film of optical thickness $\lambda/4$ but of refractive index greater than that of the glass. Then, on reflection at the top surface of the film, there will be a phase change of π between beams 1 and 2 but no phase change at the interface between film and glass because $n_1 > n$. The phase difference between reflected beams 2 and 3 is therefore π plus that due to double the optical path thickness of the film, which is also π. Hence beams 2 and 3 are in phase and will interfere constructively: the reflection is enhanced. A suitable durable material for such films is vacuum-deposited titanium dioxide, which has a refractive index of 2·8 at a wavelength of 5,000 Å. Enhanced reflection from titanium dioxide coated glass is used to make semi-transparent mirrors for optical beam-splitters; they have very low light absorption compared with metal films.

Exercise 9

1. Describe and explain the interference phenomena that can be observed when monochromatic light passes through a thin transparent film.
When a mixture of light from red and blue monochromatic sources is passed through a wedge-shaped air film between optical flats it is found that every third red fringe coincides with every fifth blue fringe. When the space between the optical flats is filled with liquid, every 33rd fringe of red light coincides with every 50th blue fringe. How may these observations be explained? (L. Anc.)

2. Explain why interference fringes are seen when monochromatic light is passed through a thin non-uniform sheet of transparent material. Describe how this effect may be used for measuring the refractive index of a liquid. If a sheet of transparent material has a stepped surface, estimate the least height of a step that could be detected if the position of a fringe can be estimated with an accuracy of 0·05 of the spacing between fringes. (Assume refractive index of material $= 1·5$ and the wavelength of light $= 5·4 \times 10^{-5}$ cm.) (L. Anc.)

3. Explain how interference fringes are produced when monochromatic light is reflected from a thin film. (Neglect the effect of multiple reflection.) Describe the experimental requirements necessary to produce (i) fringes of equal thickness and (ii) fringes of equal inclination. Discuss the fringe location in each case. (L. P.)

4. Describe and explain the phenomenon of Newton's rings. What practical application may be made of this phenomenon?
Interference bands are produced by light of wavelength 5893 Å passed

through a wedge of plastic material of refractive index 1·40. The distance between the fringes is 0·20 cm. Calculate the angle of the wedge. (L. Anc.)

5. Explain the formation of Newton's rings and describe an experiment based on the measurement of rings for determining the refractive index of a liquid.

A thin film of air is enclosed between two glass plates inclined at a small angle. When the film is viewed by reflected light of wavelength 5900 Å incident upon it, interference fringes with a separation of 1 mm are observed. Calculate the angle between the glass plates. (L. Anc.)

6. Write an essay on the blooming of optical surfaces. (L. G.)

7. Newton's rings are formed between a plane glass plate in contact with a lens of radius of curvature 2,000 cm. With sodium light passing through the air film at 30° to the normal find the diameter of the fifth black ring. Derive any formulae you use.

The experiment is repeated using a mixture of sodium light and light of another wavelength, and it is found that the eighth bright ring of the system formed by the light of the second wavelength coincides exactly with the seventh bright ring with the sodium light. Calculate the wavelength of the added light and discuss the effect on the coincidence of placing a film of liquid of refractive index 1·3 between the lens and the plate.

Assume the wavelength of the sodium light to be 5893 Å. (G. I. P.)

CHAPTER 10

Interference between Two Beams of Light: Division of Amplitude by Semi-transparent Reflector

The interference of light has led to the development of *interferometers*, which enable the optical paths of two beams to be measured relative to one another to a very high order of accuracy; and this, in turn, has led to precision methods of measuring length, refractive index, the fine structure of spectrum lines, surface contours and the inhomogeneity of transparent materials. In the development of such instruments it is frequently necessary to arrange for two (or more) beams of light to follow widely separated paths although they must arise from a single extended source in order to ensure coherence. One beam can then be modified optically in some way without disturbing the second reference beam. Subsequently the two beams, modified and unmodified, are brought together so as to traverse the same path and give an interference pattern from the measurement of which the effects of the modification can be accurately determined.

One way of obtaining two such widely separated beams from a single source is to use a semi-transparent mirror set at 45° to the axis of the incident light. The reflected beam from this 'beam-splitter' is then at 90° to the transmitted beam. Furthermore, if the mirror is carefully coated with the semi-transparent material (usually a film of aluminium deposited by evaporation of aluminium *in vacuo*), it is possible to arrange that the two separated beams have the same amplitude: the division of amplitude is equal. Such semi-transparent mirror beam-splitters could be of insignificant thickness if the aluminium were coated on a thin plastic film. Beam-splitters of this type have been used in cameras, but are not for usual interference purposes. To ensure the accurately optically flat beam-splitter needed in interferometry, a polished plane plate of glass coated with the aluminium is needed. Refraction in this plate then has to be taken into account or compensated for in some way. The pioneer exponent of such a semi-transparent mirror to obtain equal division of amplitude in interference studies was Michelson in 1881.

10.1 The Michelson Interferometer

In this instrument, light from an extended source S is incident upon an accurately plane-parallel plate A of homogeneous glass set at 45° to the axis of the incident beam (Fig. 10.1). The back-

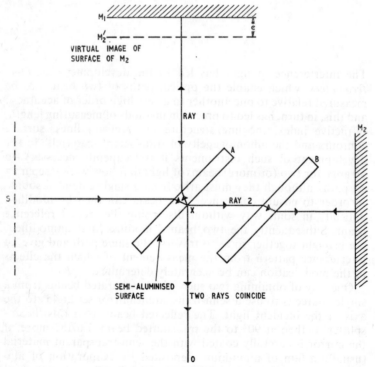

Fig. 10.1. The Michelson interferometer

surface of A (opposite to the front-surface at which the light is incident) is coated with a semi-transparent film of aluminium. After refraction in the plate A, the incident central ray is equally divided by this film into a reflected ray 1 and a transmitted ray 2, where rays 1 and 2 follow paths perpendicular to one another. The central ray 1 is reflected from the surface of the fully alumi-nised plane mirror M_1, which is mounted so that it can be adjusted until the incidence is accurately normal and is also on a precision screw so that it can be moved backwards and forwards along a line perpendicular to its plane surface. The ray 1 is, therefore, reflected back along its path of incidence, again refracted in plate A and emerges through the aluminium film along the line XO to the observer at O. Ray 2 is also reflected from a fully aluminised mirror M_2, fixed in lateral position but adjustable to be accurately perpendicular to ray 2. The ray reflected back from M_2 also reaches the aluminium film at point X, where it is reflected through 90° to traverse the path XO (the same path as ray 1 takes finally) to the observer.

It is to be noted that ray 1 undergoes two paths within the plate A. To ensure that the optical history of ray 2 is similar, the plate B—identical to A except that it has no semi-transparent film—is inserted within beam 2 and is set accurately parallel to plate A. Ray 2 thus also undergoes two paths within a glass plate. If, then, the geometrical path lengths from X to M_1 and from X to M_2 are equal, the optical paths are also equal.

If these paths are unequal, the virtual image M_2' of M_2 formed by the semi-transparent film on A will appear in front of, or behind, M_1. The overall effect of the instrument is therefore to

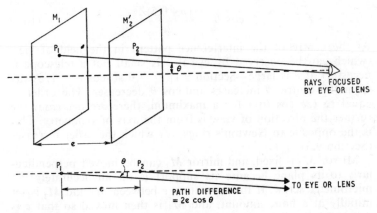

Fig. 10.2. Optical system equivalent to that of the Michelson interferometer

provide at O an interference pattern which is the same as that due to an air film enclosed between two plane surfaces and of thickness e (the difference between the path lengths between the superimposed waves at O), where $e = M_1M_2'$ (Fig. 10.2). Consequently, if M_1 and M_2 are accurately perpendicular to one another, and the surfaces of A are accurately at 45° to both M_1 and M_2, M_1 and M_2' are accurately plane-parallel; consequently, with an extended monochromatic source S of light of wavelength λ, a set of concentric circular interference fringes of equal inclination is seen in the field of view in accordance with Equation 9.1:

$$\text{optical path difference} = 2e \cos \theta$$

where $n = 1$, since air is enclosed between M_1 and M_2', and θ is the angle between the normal to mirror M_1 and a direction of view from a point not in the centre of the field (where $\theta = 0$) (Fig. 10.2).

The corresponding phase difference δ between the two wave motions at the observer is given by:

$$\delta = \frac{2\pi(2e \cos \theta)}{\lambda} \tag{10.1}$$

As the amplitudes of the two waves are equal (provided that the semi-transparent film on A is of the correct thickness), the resultant intensity is seen from Equation 7.18 to be directly proportional to $\cos^2 (\delta/2)$, which is a maximum when $\delta = 2p\pi$ and zero when $\delta = (2p+1)\pi$, p being an integer. It follows that, for a maximum,

$$\cos \theta = \frac{p\lambda}{2e} \tag{10.2}$$

and, for a minimum,

$$\cos \theta = \frac{(p+0\cdot5)\lambda}{2e} \tag{10.3}$$

At the centre of the interference pattern in the field of view (which can be observed by eye if e is small but a telescope is necessary if e is large—Section 9.1), $\theta = 0$ so $\cos \theta = 1$. Away from the centre, θ increases and $\cos \theta$ decreases. The order p, equal to $(2e \cos \theta)/\lambda$ for a maximum, therefore *decreases* the further the direction of view is from the axis of symmetry. This is the opposite to Newton's rings, for which the order increases (Section 9.3).

Mirror M_2 is fixed, and mirror M_1 can be moved perpendicularly to its plane because its support is on an accurately cut micrometer screw. If the separation e between M_1 and M_2' is set initially at a finite amount, and M_1 is then moved so that e is decreased continuously but gradually, the circular fringes seen in the field of view will progressively disappear at the centre of the field and also become more widely spaced. This is seen from Equation 10.2: λ is constant, so for a given circular bright fringe corresponding to a given order p, cos θ must increase as e is decreased. An increase of cos θ involves a decrease of θ, so the bright fringe (and similarly for the dark fringes) must move towards the centre where $\theta = 0$. Indeed, each time p in Equation 10.2 with cos $\theta = 1$ changes by unity (i.e. each time the path difference in air is decreased by $\lambda/2$), a fringe will disappear at the centre of the field of view. Furthermore, neighbouring circular fringes become more widely separated as e is made smaller until, when $e = 0$ (i.e. when the optical paths 1 and 2 are equal), the fringe pattern disappears completely and the field of view is of uniform intensity. This intensity will usually be between blackness and maximum brightness because there is a small phase change introduced on reflection at the aluminium film on A.

Localised fringes are obtained if mirror M_2 is tilted slightly away from the position where its plane surface is perpendicular to that of the plane of mirror M_1. Then M_1 and M_2' are inclined at a small angle and, if e is very small, straight optical wedge type fringes will be observed. With larger values of e, these fringes will become curved. This curvature is to be expected because a constant optical path difference is involved for any given fringe. This path difference is $2e$ cos θ; away from the centre of the field of view, θ is larger and cos θ smaller. To counteract the increase of θ, e must be bigger, and this will occur nearer the thicker parts of the wedge of air between M_1 and M_2'. The fringes are thus concave towards these thicker parts of the wedge (Fig. 10.3).

Localised fringes are frequently used in measurements with a Michelson interferometer. A main reason for this is that localised fringes become straighter as e is decreased towards zero and then

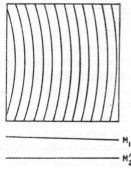

Fig. 10.3. Localised fringes in the field of view of a Michelson interferometer

curve in the opposite direction as e is increased again beyond the region where the planes of M_1 and M_2' intersect. This setting where the localised fringes become straight and few in number is comparatively easily found experimentally: it corresponds to zero, or nearly zero, path difference. At zero path difference, the order p is zero and the path difference is zero for all wavelengths. If, therefore, the extended monochromatic source is replaced by an extended source of white light, a white achromatic fringe bordered on either side by coloured fringes will be obtained near the middle of the field of view. With white light, movement of M_1 to ensure that this white achromatic fringe is central in the field of view ensures that the plane of M_2' intersects centrally that of M_1, and that the optical path lengths via M_1 and M_2 are equal to within one-tenth of a wavelength (or considerably less in a high-class Michelson interferometer). This precision setting of equal optical paths is obviously valuable in the use of the interferometer for the measurement of length.

10.2 Some Applications of the Michelson Interferometer

The Michelson interferometer is provided with small screw adjustments on the mounts of mirrors M_1 and M_2 and also with the main micrometer screw which enables mirror M_1 to be traversed a directly measured distance perpendicular to its plane.

To set up the instrument so that the circular fringes of equal inclination are obtained with a monochromatic extended source of light (usually a ground glass screen illuminated by a sodium lamp in a students' experiment), it is essential to ensure that mirrors M_1 and M_2 are accurately perpendicular to one another and each at 45° to the beam-splitter plate surface. In order to do this, it is best first to erect a large pin on the axis between the source and plate A or, alternatively, a pinhole aperture in a screen. Observation from the position O (Fig. 10.1) will then be of four images of the pin or pinhole. Two are bright images formed by reflection in the aluminium film on A, the two others are fainter because they are due to reflections in the front-surface of A, which is not aluminised; the fainter images are displaced from the bright images by an amount depending on the thickness of A. The screw adjustments to mirrors M_1 and M_2 are carefully altered until the bright images coincide, when the fainter images will also coincide. If the separations of M_1 and M_2 from X are adjusted beforehand to be nearly equal (approximately measured by an ordinary steel rule to within about 1 mm), circular fringes will appear in the field of view, and final adjustments can be made to the screws on the mounts of M_1 and M_2 to arrange these fringes centrally in the field.

With e small (i.e. when M_1 has been moved until the circular fringes are small in number), localised fringes are obtained by tilting M_2 slightly by means of one of its mount screws. These fringes are then rendered straight by traverse of M_1 perpendicular to its plane, the straight position being midway between observed positions where the fringes are curved first towards one side and then the other. White light is then substituted for the monochromatic source, and final adjustment of the position of M_1 is made until the achromatic white fringe is central in the field of view; the optical path lengths are then accurately equal, corresponding to e being zero.

10.2.1 MEASUREMENT OF THE WAVELENGTH OF SODIUM LIGHT

A simple introductory experiment is to measure the wavelength λ of sodium light in terms of the micrometer screw calibration. To do this, circular fringes are obtained with mirror M_1 beyond the image M_2' of M_2. Mirror M_1 is then moved slowly nearer to M_2', and the disappearance of circular dark fringes at the centre of the field of view is counted. If the total traverse of M_1 during the experiment as measured by the micrometer screw is l cm, whilst the number of disappearances counted is N, it follows that $N\lambda/2 = l$ because a disappearance occurs each time the path difference in air between rays 1 and 2 is decreased by $\lambda/2$. Hence λ is obtained from $2l/N$, where l is measured and N counted.

10.2.2 MEASUREMENT OF A GEOMETRICAL LENGTH

If λ is known, the same procedure as above can be used to move mirror M_1 from one position P to another Q at the extremes of a geometrical length to be measured. Counting fringes will then determine this length to within 10^{-5} cm, where the wavelength of the monochromatic source used is about 5×10^{-5} cm. The problem remains to determine when a fiducial mark on mirror M_1 is precisely located against the positions P and Q. This is best achieved by the use of a small microscope attached to the mount of mirror M_1. The cross-wires in the eyepiece of this microscope are first focused on the fine scratch mark or other means of locating position P, and M_1 is then traversed to bring the second mark against Q on to the cross-wires.

10.2.3 MEASUREMENT OF REFRACTIVE INDEX

The refractive index n of a very thin parallel-sided plane sheet of transparent material (for example, glass, mica or plastic) may be measured at a given wavelength λ. If this thin plate of geometrical thickness t is put into the path of one of the interfering beams, the light in that beam traverses the plate twice. The increase in optical path as a result of the plate over the optical path in air (of refractive index close to unity) is $2(n-1)t$. If a monochromatic source of light were used, the fringes would be displaced discontinuously on introduction of the plate, so counting of the fringe shifts would be impossible. The experiment is, therefore, undertaken with white light fringes. A satisfactory procedure is then to insert the thin plate into one of the beams, say beam 1, so that it covers about half the field of view. With monochromatic light, the mirror M_1 is moved until straight parallel localised fringes are obtained in the part of the field unobstructed by the plate. A white light source is then used in place of, or as well as, the monochromatic source, and exact equality of optical paths in the interferometer is ensured by moving M_1 until the achromatic fringe is central in the unobstructed part of the field. The number of monochromatic fringe shifts is then counted as M_1 is further moved until white light fringes are obtained in the part of the field obstructed by the thin plate.

If N is the number of fringes by which the monochromatic fringe system is displaced between the two positions where white light fringes are observed, it follows that:

$$2(n-1)t = \frac{N\lambda}{2}$$

If the wavelength λ and the thickness t are known whilst the number N is determined, n the refractive index of the material of the thin plate is found.

The refractive index of a thick parallel-sided plane plate can be found provided two identical plates are available and one is inserted in each of the beams 1 and 2. One of the plates is then turned slowly about a vertical axis through a measured small angle, and the number of fringe shifts during the rotation is measured. The change of path through the rotated plate compared with that in the stationary plate is easily related geometrically to the angle of rotation, and so the refractive index can be determined from the number of fringe shifts.

It is possible to determine the refractive index of a gas with a Michelson interferometer by admitting the gas slowly to a hollow parallel-sided glass vessel in one of the beams. However, the Jamin interferometer (Section 10.5) or the Rayleigh refractometer (Section 11.4) is preferred for this purpose.

10.2.4 STUDY OF SPECTRUM LINES

Michelson used his interferometer to study the fine structure of spectrum lines. In this connection, he defined the visibility V of the circular fringes of equal inclination obtained with an extended monochromatic source as

$$V = \frac{I_{max} - I_{min}}{I_{max} + I_{min}} \qquad (10.4)$$

where I_{max} = intensity of light at the centre of a bright fringe, and I_{min} = intensity in a neighbouring dark fringe.

A source of nominally monochromatic light, such as a discharge lamp will, in fact, not give truly monochromatic radiation of a single wavelength λ but spectral lines of a small finite width over the range λ to $\lambda + d\lambda$. This 'broadening' of the line is primarily due to Doppler effect (Section 7.8) brought about by the fact that the excited atoms producing the light are in kinetic motion; thus, any particular radiating atom may at a given time be moving in the direction of the light, away from this direction, or somewhere between. A secondary cause of the finite range $d\lambda$ is that the radiation is not truly simple harmonic because the atoms radiating are disturbed by neighbouring atoms to an extent which increases with the number of atoms per unit volume (and so with the pressure) in the discharge (an effect known as *pressure broadening*); furthermore, the radiated wave is subject to damping.

An apparently single line in the spectrum of the light from the source may also consist of two or more lines close together in wavelength. Whether or not this *fine structure* of the line is observed depends on the resolution of the spectrometer employed. Thus, there is not necessarily only a single wavelength λ present but wavelengths $\lambda_1, \lambda_2, \lambda_3 \ldots$, each of which has a small spread due to Doppler effect, pressure broadening and natural damping. The well-known example is the yellow-orange light from a sodium lamp, nominally of wavelength 5,893 Å but in fact consisting of two components of considerable intensity at 5,890 Å and 5,896 Å, differing by 1 part in 1,000 approx.

Suppose, therefore, a nominally monochromatic source actually emits light of two wavelengths λ_1 and λ_2, which are close together. Ignoring the finite width of each line and assuming for simplicity that the two radiations are, of equal intensity, consider Equation 10.3

$$2e \cos \theta = (p+0.5)\lambda$$

giving a minimum in the field of view of intensity I_{min}. There are two values of λ : λ_1 and λ_2. The value of e may be such that p is an integer for both λ_1 and λ_2 over a range of values of θ. The minima in the circular fringes of equal inclination observed in a Michelson interferometer with such a source of light will then be with I_{min} very small and the fringe visibility V high. However, at other values of e, Equation 10.3 for a given value of θ may be satisfied with $\lambda = \lambda_1$, whereas Equation 10.2

$$2e \cos = p\lambda$$

for a maximum is satisfied with $\lambda = \lambda_2$. The result will be now that the fringe visibility is poor because a dark fringe of one wavelength λ_1 will have superimposed upon it a bright fringe of wavelength λ_2.

The ratio of the wavelengths λ_1/λ_2 may be found by determining maximum and minimum visibility of fringes. An interesting students' experiment is with a sodium lamp as the source of light. This is best performed by first obtaining equality of optical paths 1 and 2 in the interferometer, ensured by the use of white light fringes. At this position, where $e = 0$, the sodium light source is inserted in place of the white light. The mirror M_1 is then moved slowly, and fringe disappearances or appearances at the centre of the field with circular fringes of equal inclination (or fringe traverses past a cross-wire in a viewing telescope eyepiece with localised fringes) are counted until the field appears to be of uniform intensity—i.e. the fringe visibility is so poor that the dark fringes are not discerned. In an experiment, 490 fringes had traversed the field centre before this condition was achieved; beyond this setting, further movement of mirror M_1 to increase e, whilst another 490 fringes were counted, restored maximum visibility.

At minimum visibility,

$$2e \cos \theta = p\lambda_1 = (p+0.5)\lambda_2$$

where light of wavelength λ_1 gives rise to a bright fringe of maximum intensity for a particular value of θ, upon which is superimposed a dark fringe of minimum intensity for the light of wavelength λ_2. As $p = 490$,

$$\frac{\lambda_1}{\lambda_2} = \frac{490.5}{490} = \frac{981}{980}$$

Therefore,

$$\frac{\lambda_1 - \lambda_2}{\lambda_2} = \frac{1}{980}$$

If, therefore, the shorter wavelength λ_2 is 5,890 Å, the larger wavelength λ_1 is greater by $\Delta\lambda = \lambda_1 - \lambda_2 = 5,890/980 = 6$ Å, i.e. $\lambda_1 = 5,896$ Å.

Michelson, with his vast experience of interferometry, was adept at obtaining settings of fringe visibility and indeed was able to obtain the wavelength spread of the λ to $\lambda + d\lambda$ of spectrum lines, as well as investigate fine structure. The Michelson interferometer is not, however, an easy instrument to use for these purposes as discernment of the settings of maximum and minimum fringe visibilities involves considerable error of observation. The Fabry–Perot interferometer has replaced the Michelson instrument in present-day practice for examining the fine structure of spectra.

10.3 Comparison of the Standard Metre with the Wavelength of Light by means of a Modified Michelson Interferometer

The standard of length, the metre, is accepted internationally to be that between fiduciary marks on a uniform bar of platinum–iridium maintained at constant temperature and preserved at Sèvres, near Paris. It is manifestly important to determine wavelengths of light as exactly as possible in terms of this accepted standard. Furthermore, once the wavelength of a substantially monochromatic radiation from a specified source is thus found, the standard metre can be reproduced anywhere.

This determination was first carried out in 1892–3 by Michelson and Benoît. They used a cadmium lamp as the source of radiation, with particular reference to the red line in its spectrum. More recent work has employed sources giving light having a smaller wavelength spread than cadmium and other interferometers than the Michelson (Chapter 4, Volume 2). Though Michelson and Benoît's original work has been superseded, its importance is such as to merit study in some detail.

It would appear, on first thoughts, that, with a cadmium lamp as the source, fringe traverses across the field of view could be counted as the movable mirror M_1 of a Michelson interferometer was traversed from one end of the standard metre to the other. This is virtually out of the realm of practice, however, because fringes with an optical path difference of 1 m would be difficult to discern and a formidable number of fringes would need to be counted during the traverse of the movable mirror.

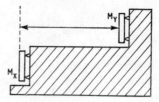

Fig. 10.4. The Michelson–Benoît etalon (the plane of mirror M_x can be made parallel to that of M_y by means of adjusting screws)

Michelson and Benoît therefore constructed nine etalons, each of the form shown in Fig. 10.4; the longest had a separation between the front surfaces of mirrors M_x and M_y of 10 cm, the next 5 cm, and so on, with each succeeding etalon approximately half the length of its predecessor, down to the smallest of 0·39 mm. These lengths were, of course, nominal in that they were determined by precise engineering measurement but did not have to be exactly related to the standard metre. The Michelson interferometer, slightly modified, was then used first to determine the

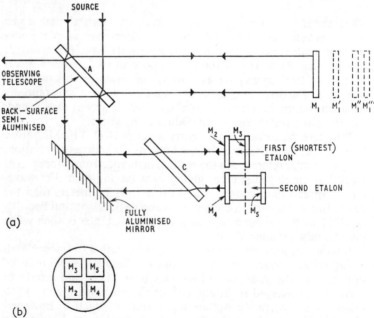

(a)

(b)

Fig. 10.5. (a) Arrangement of Michelson and Benoît for measuring with an interferometer the number of wavelengths in the shortest etalon, and for comparing the lengths of the shortest and second shortest etalons; (b) field of view

number of wavelengths in the distance between the mirrors of the shortest etalon. The arrangement of Fig. 10.5 was employed to determine this number for the 'monochromatic' red, green and blue lines of the cadmium discharge lamp spectrum. Then the etalons were compared in pairs, i.e. the shortest etalon with the next shortest, and so on, to give eventually the number of wavelengths in the longest 10 cm etalon; this was finally compared with the standard metre.

In the determination [Fig. 10.5(a)], M_1 is the movable mirror on an accurate micrometer screw of the Michelson interferometer, whilst M_2 and M_3 are the mirrors of the shortest etalon. Within this diagram are also shown M_4 and M_5, the mirrors of the second etalon (of length nominally twice that of the first). Each of the square plane mirrors of these etalons is of an area just less than a quarter of that of mirror M_1, so that the field of view in the interferometer is divided into four square sections [Fig. 10.5(b)] in which the interference fringes observed correspond to effects obtained in the mirrors M_2, M_3, M_4 and M_5 respectively. The procedure is as follows.

1. Localised fringes are used and obtained by tilting slightly the mirrors of the first etalon. The field is viewed by a telescope with an eyepiece furnished with cross-wires.

2. White light fringes are obtained within the field of view concerned with the surfaces of mirrors M_2 and M_4. To do this, mirror M_1 is moved until the optical distances to the observer from M_1 and M_2 are equal. Then the second etalon is moved so that its mirror M_4 is coplanar with mirror M_2 of the first etalon, when white light fringes will be spread across M_2 and M_4 in the field of view.

3. The light from one of the cadmium source lines is substituted for the white light, and mirror M_1 is moved slowly, the number of fringes which cross the centre of the field of view (i.e. the cross-wire of the telescope eyepiece) being counted until the optical paths to the observer from M_1 and M_3 are equal. To ensure final equality of path, with M_1 at M_1', a check with white light is essential. The fractional part of the cadmium light fringe displacement which will give exact equality of optical paths is determined by rotating slightly through a measured angle the compensating plate C. The number of wavelengths equal to the separation of the mirrors M_2 and M_3 is thus determined for each of the three lines in the cadmium spectrum. Let the number of wavelengths equal to the length of the first etalon be n_1 for a given cadmium line; n_1 is not necessarily an integer.

4. The shortest etalon is moved through its own length so that its mirror M_2 is in the position originally occupied by its mirror M_3. White light fringes are observed across M_2 when this setting is correct. Then M_1 is moved to M_1'', the position where white light fringes are observed across M_3. The number of wavelengths equal to the distance between the original position of M_1 and M_1' is then $2n_1$.

5. Cadmium fringes are counted while M_1 is moved from M_1'' to M_1''' until white light fringes are observed across M_5. This movement of M_1'' to M_1''' will be very small because the second etalon is nominally twice the length of the first etalon. The number of cadmium fringes counted will, therefore, be a small number c. If n_2 is the number of wavelengths equal to the length of the second etalon, it follows that:

$$n_2 = 2n_1 + c$$

This result relates the length of the second etalon to that of the first etalon.

The process of comparing etalon lengths in terms of numbers of wavelengths they contain is then repeated for the second and third etalons, the third and fourth etalons and, finally, the eighth and ninth etalons. Thus, the length of the nominally 10 cm etalon is known in terms of the number of wavelengths it contains.

6. The final stage is the comparison of the longest 10 cm etalon with the standard metre. A pointer P on the mirror M_x of

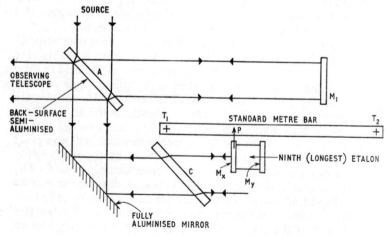

Fig. 10.6. Comparison of the 10 cm etalon with the standard metre

this etalon is set in coincidence with the end mark T_1 of the standard metre, a microscope being used for observation of this coincidence. The movable interferometer mirror M_1 is set coplanar with M_y, as determined by white light fringes (Fig. 10.6). The etalon is moved through its own length until the white light fringes are central within mirror M_x. The mirror M_1 is moved until white light fringes reappear in M_y, the etalon is moved bodily again, and the process is repeated until after ten displacements the pointer attached to M_x appears at the other end mark T_2 on the standard metre. Cadmium light fringes are then counted as M_x is moved to ensure precise coincidence of the pointer with end mark T_2, as observed under a microscope.

The total number of wavelengths in the standard metre is thus determined.

In the determination, it is important to note that errors are not cumulative. The fractional part of a fringe measured in determining the number of wavelengths in the first etalon is used to make sure of the whole number of wavelengths in the second etalon, and so on in comparing the other pairs of etalons. The error involved in measuring the length of the longest etalon in terms of wavelength is less than that which arises in judging the coincidence of the pointer P with the end marks on the standard metre.

Accepting the standard metre as the international unit of length, the wavelengths *in vacuo* of the three lines in the cadmium spectrum are determined to be: red, 6,438·4722 Å; green, 5,085·8240 Å; blue, 4,799·9107 Å. Alternatively, it may be said that

$$1 \text{ m} = \frac{10^{10}\lambda}{6,438\cdot4722} = 1,533,163\cdot5\lambda$$

where λ is the wavelength of the cadmium red line.

10.4 The Twyman and Green Interferometer: Testing of Prisms and Lenses

In the manufacture of high-quality prisms and lenses, it is important to be able to investigate the precision of the working of the surfaces and the homogeneity of refractive index of the optical glass used. This can be done by investigating the aberrations introduced into a wavefront of monochromatic light which traverses the optical component; the most sensitive method is by means of interferometry. The Michelson interferometer can be

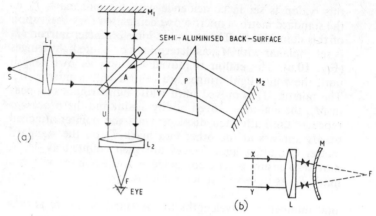

Fig. 10.7. *The Twyman–Green interferometer: (a) testing a prism;
(b) testing a lens*

used to some extent for this purpose but it is better modified to the instrument of Twyman and Green, both formerly of Adam Hilger Ltd, specialists in high-precision optical working. The modifications are modest: the incident monochromatic light on to the beam-splitter mirror A, which is set accurately at 45° and has its back-surface semi-aluminised, is in the form of an accurately collimated broad beam from a well-corrected objective lens L_1 having a point source S at its focus. The wavefront of the light beam which undergoes amplitude division at A is, therefore, plane (Fig. 10.7). The mirror M_2 on its mount is movable over the horizontal flat steel bed of the interferometer so that it can be set to receive light from various directions. For example, in Fig. 10.7, M_2 is moved to be normal to light which has undergone refraction by the prism P. The interfering light beams are brought to the eye at the focus of a second corrected objective lens L_2.

With prism P absent and M_2 moved to be parallel to the incident plane wavefront XY, the two re-uniting wavefronts in such a position as UV are plane. The well-corrected lens L_2 will bring the corresponding incident parallel light to a focus at F where the eye is placed. Across the whole field of view there is a constant phase difference between the two superposed beams and not, as in the Michelson interferometer, a variation with angle: the field will, therefore, be uniformly illuminated at maximum brightness if the path difference between the interfering beams is zero or an integral number of wavelengths, and at minimum brightness when the difference is an odd integral number of wavelengths. Note that the circular interference fringes of equal inclination seen in the Michelson interferometer using an extended source are not present.

To examine a prism P, it is placed in the parallel beam having a plane wavefront such as XY. The mirror M_2 is moved so that the light emerging from P is normally incident. If the prism P were perfect (i.e. with optically plane refracting surfaces and of constant refractive index through its material), the returned wavefront would also be plane, and the field of view at the focus of lens L_2 would still be uniformly illuminated. If, however, the prism were defective (for example, if the refracting surfaces were not accurately plane and/or the refractive index of its material were not uniform), the wavefront returned by the mirror M_2 would be deformed relative to the incident plane wavefront at XY. Consequently, plane and non-planar wavefronts would reach the lens L_2, and correspondingly there would be a variation of illumination across the field of view. The interference fringes now seen may be regarded as contour lines of the deformed wavefront, with neighbouring fringes one wavelength apart. The experienced optical worker can mark the region of the prism which gives rise to such contour fringes and subsequently locally polish the surfaces until the Twyman and Green interferometer shows a uniformly bright field of view, corresponding to optical perfection of the prism.

Similarly, the perfection of a converging lens L may be examined and surface figuring undertaken as required. The plane mirror M_2 is replaced by the accurately spherical convex mirror M [Fig. 10.7(b)]. If the lens L were perfect, it would refract the incident light with a plane wavefront such as XY to a point focus at F. If F is also the centre of curvature of the spherical convex mirror M, the light incident at M is returned along its path of incidence to give interfering plane wavefronts incident on the lens L_2 and a field of view of uniform illumination at the eye. On the other hand, if L_2 has imperfections due to incorrect surface figuring, inhomogeneity of refractive index, or inadequate correction of its aberrations, the returning wavefront will be deformed; it will interfere with the plane wavefront to give revealing interference fringes in the field of view, again related to contours of imperfection in the lens.

10.5 The Jamin Interferometer: Measurement of the Refractive Index of a Gas

In this instrument, two separated parallel beams of light from a single extended monochromatic source are produced. Two separated plane-parallel plates of glass of exactly the same thickness

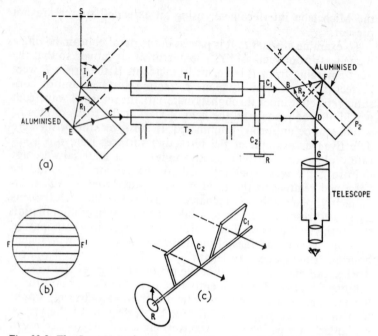

Fig. 10.8. *The Jamin interferometer: (a) plan view; (b) horizontal fringes in the field of view of the telescope eyepiece; (c) details of the compensator*

and cut from a single plane-parallel piece of high-quality optical glass about 2 cm thick are used. These plates are fully aluminised on their back-surfaces, i.e. opposite to the faces at which the light is incident. The extended source of light S, sometimes with a cylindrical lens to ensure uniform illumination of the first plate P_1 [Fig. 10.8(a)], is amplitude divided into two parallel beams AB and CD, where AB arises from reflection at the front-surface of P_1, and CD after refraction along AE and reflection along EC. These parallel beams AB and CD enter the second plate P_2, where the combination of refraction and reflection of AB via BFD and reflection of CD at D brings the two beams together; they are thus superposed along DG and enter a telescope.

With angles of refraction R_1 in plate P_1 and R_2 in plate P_2, as shown in Fig. 10.8, the optical path difference between the superposed rays entering the telescope along DG is

$$n[(AE+EC)-(BF+FD)] = n(2t \cos R_1 - 2t \cos R_2)$$

where n is the refractive index of the glass from which both plates are cut, and t is the thickness of plate P_1 which is the same as that of plate P_2. Further, with an angle of incidence of I_1 of ray SA at plate P_1, $n = (\sin I_1)/(\sin R_1)$ from Snell's law.

If the plates P_1 and P_2 are parallel to one another, $R_1 = R_2$ so this optical path difference is zero. In addition, though the angle of incidence I_1 will have various values other than that shown for the central ray SA because the source is extended, yet R_1 will always equal R_2. Consequently, the field of view in the telescope is of uniform brightness.

If one of the plates is tilted slightly about a horizontal axis parallel to its surface (for example, if P_2 is tilted about the axis X through a small angle θ), a series of nearly straight horizontal fringes will appear in the field of view of the telescope [Fig. 10.8(b)] and will be closer together the larger the angle of tilt θ. To explain this, consider the parallel rays AB and CD in the space between the plates P_1 and P_2. In the horizontal plane defined by the plane of the paper in Fig. 10.8, these rays have zero path difference. With a tilt given to plate P_2, however, rays displaced above or below AB and CD in the horizontal plane of the paper will have unequal optical paths in plate P_2 and so will give rise to bright or dark fringes above and below the central line of the field of view, depending upon whether the optical path difference due to the tilt is an integral multiple of the wavelength λ or an odd integral multiple of $\lambda/2$ respectively.

To measure the refractive index of a gas at a given pressure (measured by a suitable manometer) and recorded temperature, tubes T_1 and T_2 (of equal length and with end faces closed by parallel plane optical flats) are introduced into beams AB and CD respectively. Both tubes are first evacuated to a pressure of 1 mm Hg or less, then the gas concerned is admitted slowly to one of the tubes, say T_2, and the number of fringes such as FF' [Fig. 10.8(b)] which are displaced across the field of view is counted until the final required gas pressure is attained.

As counting numerous fringe passages in this way is tedious, and so is the gradual introduction of the gas, better practice is to use the compensator consisting of plates C_1 and C_2 [Fig. 10.8(c)], with C_1 in beam AB and C_2 in beam CD. The angle between these plates can be varied and pre-set, and the angle of setting of the combination can be recorded by the divided circle attached to the rotation head R. As this pair of compensator plates is rotated as a whole, an optical path difference is introduced between beams AB and CD. For a given angle of setting between C_1 and C_2, the angle of rotation (measured on the divided circle) imparted to the combination of C_1 and C_2 as a whole by head R can be plotted against the optical path difference as determined by fringe shifts across the field of view. The sensitivity of this compensator can be varied by changing the pre-set angle between C_1 and C_2: the smaller this angle, the larger the rotation necessary to introduce a given optical path difference, so the more sensitive the compensator.

Now to determine the refractive index of a fluid, tubes T_1 and T_2 are evacuated to a pressure below 1 mm Hg, a white light source is used and the compensator (with a given angle between C_1 and C_2) is rotated to bring the central achromatic fringe on to the cross-wires of the telescope eyepiece. The fluid (gas at a measured pressure or transparent liquid) is then introduced into one of the tubes, say T_2, and the compensator rotated to restore the central achromatic fringe to the cross-wires. The optical path difference due to the introduction of the fluid is then known from the calibration graph of angle of rotation of the compensator (for a given angle between C_1 and C_2) plotted against the number of fringe shifts for monochromatic light of a given wavelength λ. If l is the length of either tube T_1 or T_2, this determined optical path difference given by $p\lambda$ (p being the number of bright or dark fringe shifts) is related to the refractive index n by the equation:

$$p\lambda = (n-1)l$$

The Jamin interferometer for the measurement of the refractive index of fluids has been superseded by the Rayleigh instrument which, though also an interferometer, is most appropriately described under diffraction in Chapter 11 because it utilises a separated pair of parallel slits. The Rayleigh instrument is more accurate and does not demand the accurately worked homogeneous plates P_1 and P_2 necessary for the Jamin.

The important subject of multiple-beam interferometry is dealt with in Volume 2.

Exercise 10

1. Distinguish between wave-front division and amplitude division in interference systems, and give a practical example of each in two-beam interference systems. Discuss the localisation of the fringes in the examples quoted. (L. P.)

2. Explain the formation and localisation of circular interference fringe systems using monochromatic light transmitted through (a) a Newton's rings apparatus and (b) a Michelson interferometer. Point out any differences and similarities between these fringe systems and explain the effect on each of replacing the monochromatic source by white light. (L. P.)

3. Derive the equation $ml = 2 \mu t \cos r$ for optical interference fringes arising from films, where m is integral, the wavelength of the illumination is l, and r is the angle of refraction in a film of thickness t and of material of refractive index μ. Describe and sketch the experimental arrangements necessary to show three types of interference fringes according as m varies with l, t, or r and mention an application of one type. (L. G.)

4. Explain the formation of fringes in a Michelson interferometer illuminated by (a) monochromatic light, (b) white light. If one mirror is slowly moved back with velocity v calculate directly the number of monochromatic fringes passing a point in the field of view in one second. Show that the same result is obtained by a treatment which considers the beats formed between the incident light and the reflected light whose frequency has been modified by the Doppler effect at the moving mirror. (L. P.)

5. Describe the Michelson Interferometer and show how it may be used to produce (a) fringes of constant thickness and (b) fringes of constant inclination.

It is desired to measured small rotations of one of the mirrors of such an interferometer and it is found possible to locate a fringe of constant thickness to within one tenth of a fringe interval. If one cm width of mirror is used and the light is of wavelength 5461 Å to what precision can the orientation of the mirror be measured? (L. G.)

6. Show that the condition for reinforcement in the light transmitted across an air gap between two plane parallel partially reflecting surfaces is $2t \cos \alpha = n\lambda$ where the symbols have their usual meanings. Explain the formation of interference fringes in an interferometer which makes use of this condition, and describe one application of such an instrument. (L. Anc.)

7. Write an account of the design and use of an interferometer intended for the testing of the surface contours and homogeneity of the glass of prisms and lenses.

8. Describe an optical interferometer which can be used to determine the difference $\Delta\lambda$ in wavelength between two very close spectral lines. Show how $\Delta\lambda$ is measured and discuss the instrumental factors which set a lower limit to its measurable value. (L. G.)

9. Discuss the shape and localisation of the fringes observed in a Michelson interferometer. Explain the importance of this instrument in the history of interferometry. (L. P.)

10. Explain the principle of the Michelson interferometer and describe one modification used in testing optical components.

The Michelson interferometer is adjusted to give fringes with white light. Sodium light is then substituted and the movable mirror is slowly moved. When it has moved about 15×10^{-3} cm, the fringes are found to have disappeared but they return to full contrast when it is moved a further 15×10^{-3} cm in the same direction. Explain this effect and calculate the approximate separation of the two sodium D lines given that their mean wavelength is 5893 Å.   (G. I. P.)

CHAPTER 11

Diffraction of Light: Introductory Theory and Practice of Fraunhofer Diffraction

As emphasised previously, there is not always the clear distinction between diffraction and interference in practice or in theoretical treatment that an introductory account indicates. Both depend on the principle of superposition (Section 7.3): interference is, in general, concerned with two or more coherent beams of light from two or more apertures involving either wavefront division or amplitude division or both; diffraction is concerned with the interactions between light traversing the various regions of any one aperture. Furthermore, the theory of diffraction and of interference is mathematically complex if a three-dimensional picture of phenomena is described, so simplifying assumptions (sometimes closely realisable in practical experiments) are necessary to avoid the full treatment of diffraction by the use of Kirchhoff's mathematical formulation of the simple Huygens' principle. In this volume, these simplifications are to be made and the consequences of these simplifications pointed out; the fuller theory is deferred to Volume 2 on the basis that, later in a course on optics, the reader will have become familiar with vector calculus.

The occurrence of shadows consequent upon placing an opaque obstacle between a source of light and a screen is a matter of everyday observation. The reader will be familiar with the ideas of umbra and penumbra in shadows. Such observations are simply explained on the assumption that light is propagated in straight lines.

More detailed observation of the edge of a shadow produced, for example, by a straight edge shows that, in fact, there is not in the shadow an abrupt discontinuity between the light and dark regions: over a small region on either side of the superficially sharp-edged shadow, there is an alternation of light and dark regions in narrow bands. Again, if a circular aperture in a shield is situated between a small source of light and a screen, a circular patch of light appears on the screen. If the aperture is comparatively large, the departure from a well-defined edge to the uniform circular light patch is frequently too small to be important. If, however, the aperture is small, the alternations of light and shade become the most significant feature of the appearance at the screen.

These departures from the effects explicable on a geometrical basis of rectilinear propagation are explained by resort to the wave theory of light. The effects produced by light are due to wavefronts, and the way in which these wavefronts are modified on traversing apertures or encountering obstacles involves the study of diffraction.

11.1 Fraunhofer and Fresnel Diffraction

In the classification of diffraction phenomena, the terms *Fraunhofer diffraction* and *Fresnel diffraction* are used. To distinguish between these two in an introductory account, suppose a point source of light at P illuminates a screen S, and between P and S there is an aperture or obstacle (or a number of such) which form the means of diffraction D. If the distances PD and DS are very large compared with the dimensions of D and with the wavelength of the radiation, Fraunhofer diffraction* is concerned. If, on the other hand, the distances PD and DS cannot be regarded as very large compared with the dimensions of D or the wavelength, Fresnel diffraction is involved.

An example of Fraunhofer diffraction which is of frequent interest is where the light incident at D is collimated (for example, by a lens system), so the source is in effect at an infinite distance, the wavefront encountering D is plane, and the light after diffraction is brought to a focus at the receiving screen by a lens or system of lenses. In Fresnel diffraction, the distances concerned

---

*Fraunhofer diffraction does not necessarily involve large distances: see Chapter 6 of Volume 2.

are finite and usually, but not necessarily, lenses are not involved. Consideration of Fresnel diffraction is deferred to Volume 2.

11.2 Fraunhofer Diffraction at a Single Slit

The slit is taken to be a rectangular aperture in an opaque shield and to have a width a which is small compared with its length. A collimated beam of monochromatic light of wavelength λ, produced by a point source S at the focus of a spherical lens L_1, is incident normally upon this slit; the cross-section of the beam is considerably larger than that of the slit. Beyond the slit is a second converging lens L_2, which focuses the emergent light at a screen Z [Fig. 11.1(a)]. The central point on this screen Z where it is intersected by the axis of symmetry of the arrangement is at

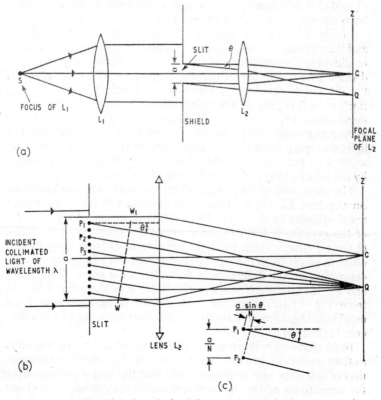

(a)

(b)

(c)

Fig. 11.1. Fraunhofer diffraction at a single slit

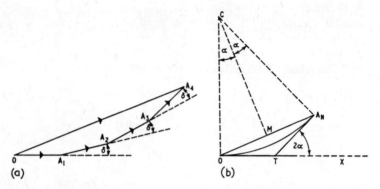

Fig. 11.2. Vector diagrams relating to the calculation of the amplitude of the light disturbance at a point on the pattern due to Fraunhofer diffraction at a slit

C, and Q is any point on the screen which receives light emerging from the slit at an angle θ to this axis.

In Fig. 11.1(b), the width of the slit is magnified for clarity and the collimator lens L_1 is omitted. So that an expression can be reached for the amplitude of the light disturbance at Q, the slit is divided into a number of narrow strips parallel to the edges of the slit and of equal width with centres at points P_1, P_2, P_3, ... Each strip is regarded as a source of secondary wavelets: each of these secondary sources in the plane of the slit will emit waves having the same period, phase and amplitude of vibration. The light emerging from these contiguous strips in a direction at an angle θ to the normal to the plane of the slit will be converged by the lens L_2 to Q in the focal plane of the lens.

The wavefront of these parallel rays at angle θ before incidence on the lens L_2 will be plane and represented by WW_1 drawn perpendicular to the rays. The optical distances along any one of the rays from WW_1 in the plane of the diagram to the point Q are all equal. Thus, the light arriving at Q from strip P_1 will differ in phase from the light arriving from strip P_2 by an amount depending on the difference in paths from the plane WW_1 to P_1 and P_2. If there are N strips in the width a of the strip, it is seen from Fig. 11.1(c) that the path difference between rays from neighbouring strips is $(a \sin \theta)/N$. The corresponding phase difference is $(2\pi a \sin \theta)/N\lambda$.

In an elementary treatment, it is assumed that the strip elements radiate uniformly in all forward directions (i.e. there is no variation of intensity with obliquity), and that the vectors representing the amplitudes of the disturbances which they produce at Q are all equal in length (which is tantamount to assuming that the differences in path lengths from P_1, P_2, P_3, ... to Q are small enough to ignore fall-off of amplitude with distance). A simple vector diagram can then be drawn to find the resultant amplitude at Q due to the N sources, each producing the same amplitude of disturbance at Q but where neighbouring sources create disturbances at Q which differ in phase by $\delta = (2\pi a \sin \theta)/N\lambda$. Thus, vector OA_1 in Fig. 11.2(a) is drawn to represent in length the amplitude at Q due to strip P_1; vector A_1A_2 at angle δ to OA_1 then represents the amplitude at Q due to strip P_2; correspondingly, A_2A_3 at 2δ to OA_1 represents the contribution of P_3, and A_3A_4 at 3δ to OA_1 the contribution of P_4 (assuming at first that only four strip elements are chosen). In this vector diagram, $OA_1 = A_1A_2 = A_2A_3 = A_3A_4$. The resultant is then represented in magnitude and direction by the vector OA_4.

It is obviously preferable to choose a large number N of strip elements. In the limiting case where N tends to infinity, the vectors OA_1, A_1A_2 and so on become infinitesimally short and lie along the arc of a circle between O and A_N, the termination of the vector representing the contribution of the final Nth slit [Fig. 11.2(b)]. The angle between the tangent OT to this arc at O and the tangent TA_N at A_N is then the phase angle between disturbances arriving at Q from the extreme edges of the slit. This angle 2α is obviously $N\delta$, where δ is $(2\pi a \sin \theta)/N\lambda$, and is, therefore, $(2\pi a \sin \theta)/\lambda$.

In Fig. 11.2(b), the length of the chord OA_N therefore represents the resultant amplitude at Q due to the N disturbances from the strip elements, where N is a very large number. If these N disturbances had all acted in phase at Q, the resultant amplitude would be represented by the length of the arc OA_N. They would all act in phase at $\theta = 0$, corresponding to the central point C on the screen Z in Fig. 11.1, It therefore follows that the amplitude of the resultant R_θ at a point Q on the screen due to light which has been diffracted through an angle θ by the slit is related to the amplitude R_0 of the resultant at the central point where $\theta = 0$ by:

$$\frac{R_\theta}{R_0} = \frac{\text{length of chord } OA_N}{\text{length of arc } OA_N}$$

In Fig. 11.2(b) it is seen that

$$\angle A_NTX = \angle OCA_N = 2\alpha$$
$$\text{chord } OA_N = 2OC \sin \alpha$$

and

$$\text{arc } OA_N = 2OC\alpha$$

Therefore,

$$\frac{R_\theta}{R_0} = \frac{\sin \alpha}{\alpha} \tag{11.1}$$

where $\alpha = (\pi a \sin \theta)/\lambda$. As the intensity of the light is proportional to the square of the amplitude,

$$\frac{I_\theta}{I_0} = \frac{\sin^2 \alpha}{\alpha^2} \tag{11.2}$$

where I_0 is the maximum intensity at the centre of the diffraction pattern where $\theta = 0$, and I_θ is the intensity at a point in this pattern corresponding to light which has been diffracted by the slit through an angle θ.

An alternative mathematical treatment of Fraunhofer diffraction is now considered. In this analytical treatment, as in the foregoing account, two simplifying assumptions are necessary.

1. The amplitude of the disturbance at any point on the screen due to light from a small element of the slit will depend on the optical distance from the element to the point. The optical distances to a given point from the various slit elements are assumed to be sufficiently nearly equal for this amplitude to be assumed to be the same irrespective of the element concerned.

2. Light rays arriving at a given point on the screen will come from different directions depending upon the element of the slit which gives rise to them. This variation of angle is ignored (i.e. the obliquity factor is ignored).

In Fig. 11.3, let O be the centre of the slit of width a. The slit is divided into parallel equal elements of width ds. The area of a given element will be $l\,ds$, where l is the length of the slit, perpendicular to the plane of the diagram. Any point within $l\,ds$

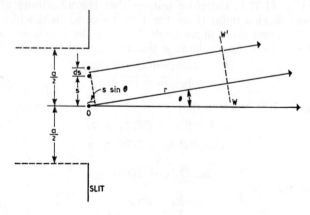

Fig. 11.3. A mathematical treatment of Fraunhofer diffraction at a slit

will emit spherical wavelets when the slit is illuminated by normally incident collimated light. The envelope of these secondary wavelets coming from a strip element of area $l \, ds$ will be a cylinder having the element as axis. The displacement $d\psi$ in a cylindrical wavefront at a distance r from the central strip element at O will be given by Equation 7.13 as

$$d\psi = \frac{A}{\sqrt{r}} \exp j(\omega t - kr) \tag{11.3}$$

where ω is the pulsatance ($\omega = 2\pi f$, where f is the frequency of the light vibrations), A is the amplitude at unit distance from the element, and $k = 2\pi/\lambda$. As the slit is uniformly illuminated by normally incident collimated light of wavelength λ, A is proportional to $l \, ds$. Further, in finding the displacement amplitude at a given point on the screen, it is assumed in accordance with (2) above that variations of r are insignificant. So Equation 11.3 may be written

$$d\psi = c_1 \, ds \exp j(\omega t - kr) \tag{11.4}$$

where c_1 is a constant.

For light leaving the slit at an angle θ to the normal to the slit, the distance to the perpendicular wavefront WW' will be r for the central element at O, and $r - s \sin \theta$ for an element ds at a distance s above O. This change from r to $r - s \sin \theta$ will affect the phase of the disturbance at WW' even though it is assumed that it does not affect the amplitude. The appropriate expression for the displacement $d\psi$ at the wavefront due to any element ds is, therefore,

$$d\psi = c_1 \, ds \exp j[\omega t - k(r - s \sin \theta)]$$

The resultant displacement due to all the elements of width ds in the slit of width a is therefore obtained by integration over the limits $s = -a/2$ to $s = +a/2$. Therefore,

$$\psi = c_1 \int_{-a/2}^{a/2} \exp j[\omega t - k(r - s \sin \theta)] \, ds$$

$$= c_1 \exp j(\omega t - kr) \int_{-a/2}^{a/2} \exp(jks \sin \theta) \, ds \tag{11.5}$$

$$= c_1 \exp j(\omega t - kr) \left[\frac{\exp(jks \sin \theta)}{jk \sin \theta} \right]_{-a/2}^{a/2}$$

$$= c_1 \exp j(\omega t - kr) \frac{\exp[(jka \sin \theta)/2] - \exp[(-jka \sin \theta)/2]}{jk \sin \theta}$$

$$= 2c_1 \exp j(\omega t - kr) \frac{\sin[(ka \sin \theta)/2]}{k \sin \theta}$$

The amplitude $R_{\theta W}$ of the disturbance at a wavefront, such as WW' a distance r from the slit, due to light diffracted by the slit through an angle θ is, therefore, given by the equation:

$$R_{\theta W} = 2c_1 \frac{\sin\left[(ka \sin \theta)/2\right]}{k \sin \theta}$$

To find the amplitude R_{0W} at this wavefront due to light passing along the central axis normal to the slit (i.e. for $\theta = 0$), consider that Equation 11.5 becomes:

$$\psi = c_1 \exp j(\omega t - kr) \int_{-a/2}^{a/2} ds = c_1 a \exp j(\omega t - kr)$$

Therefore, the ratio

$$\frac{R_{\theta W}}{R_{0W}} = \frac{2c_1 \sin[(ka \sin \theta)/2]}{(k \sin \theta)c_1 a}$$

This ratio of amplitudes at the wavefront will be the same as that at the screen after focusing by the lens. Therefore,

$$\frac{R_\theta}{R_0} = \frac{\sin\left[(ka \sin \theta)/2\right]}{(ka \sin \theta)/2}$$

Substituting $k = 2\pi/\lambda$,

$$\frac{R_\theta}{R_0} = \frac{\sin[(\pi a \sin \theta)/\lambda)]}{(\pi a \sin \theta)/\lambda}$$

where R_θ is the amplitude of the disturbance at a point on the screen due to light diffracted by the slit through an angle θ, and R_0 is the amplitude at the centre of the screen where it is intersected by the normal to the plane of the slit. Putting $\alpha = (\pi a \sin \theta)/\lambda$,

$$\frac{R_\theta}{R_0} = \frac{\sin \alpha}{\alpha}$$

and the ratio of the intensities is

$$\frac{I_\theta}{I_0} = \frac{\sin^2 \alpha}{\alpha^2}$$

These results agree with Equations 11.1 and 11.2.

In Fig. 11.4 is plotted the light intensity I_θ in the diffraction pattern at the screen against the angle α. This will represent the variation in intensity in traversing the screen along an axis perpendicular to that of the slit. Fig. 11.4 is plotted from the calculated values shown in Table 11.1, where the maximum amplitude at the screen centre at C is taken to be unity.

Table 11.1. Values of $(\sin \alpha)/\alpha$ and $(\sin^2\alpha)/\alpha^2$ for Different Values of α

α (rad)	0	$\pi/4$	$\pi/2$	$3\pi/4$	π	$5\pi/4$	$3\pi/2$	2π	$5\pi/2$	3π
$(\sin \alpha)/\alpha$	1	0·9	0·61	0·3	0	−0·18	−0·213	0	0·128	0
$(\sin^2 \alpha)/\alpha$	1	0·81	0·41	0·09	0	0·033	0·046	0	0·016	0

Note the central maximum flanked on either side by minima, at which the light intensity is zero and where there are hence dark fringes. The positions of the minima are determined by the equation $\alpha = p\pi$ (or $-p\pi$); i.e. dark fringes occur at angles θ given by

$$\alpha = \frac{\pi a \sin \theta}{\lambda} = p\pi$$

where p is a positive integer to one side of the central maximum and a negative integer on the other side. Therefore,

$$\sin \theta = \frac{p\lambda}{a} \qquad (11.6)$$

The first minima encountered on moving away from the central

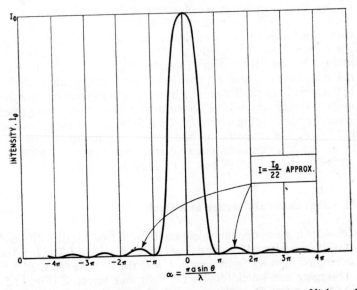

Fig. 11.4. Fraunhofer diffraction at a single slit: plot of intensity of light at the screen against angle α

maximum will therefore occur when $p = 1$ (or -1). The corresponding angle of diffraction is given by:

$$\theta_1 = \sin^{-1}\frac{\lambda}{a}$$

As a is usually considerably larger than λ, $\theta_1 = \sin \theta_1$ so

$$\theta_1 = \frac{\lambda}{a} \tag{11.7}$$

The positions of the maxima *do not* occur exactly midway between the minima, as is shown in Fig. 11.4. These locations can also be found by putting

$$\frac{dR_\theta}{d\alpha} = \frac{d}{d\alpha}\left(\frac{\sin \alpha}{\alpha}\right) = 0$$

Therefore,

$$\frac{\alpha \cos \alpha - \sin \alpha}{\alpha^2} = 0$$

i.e.

$$\alpha = \tan \alpha$$

This equation is best solved graphically by finding the points of intersection of the plots of $y = \alpha$ and $y = \tan \alpha$. The values found are $\alpha = 0$, $1 \cdot 43\pi$, $2 \cdot 46\pi$, $3 \cdot 47\pi$, and so on. Apart from the central maximum, these values are displaced slightly towards the centre from the positions $1 \cdot 5\pi$, $2 \cdot 5\pi$, $3 \cdot 5\pi$, ..., which are midway between the minima. However, the intensities in the maxima may be calculated approximately by assuming that they *are* at these midway positions. The intensities in the maxima are then unity at the centre, followed towards one side by successive values given by substituting $\alpha = 1 \cdot 5\pi$, $2 \cdot 5\pi$, $3 \cdot 5\pi$, ... in the expression $(\sin^2 \alpha)/\alpha^2$. The intensities of the central maximum and the first three maxima towards one side are consequently 1, $1/(3\pi/2)^2$, $1/(5\pi/2)^2$ and $1/(7\pi/2)^2$; i.e., 1, 1/22, 1/62 and 1/121.

11.3 Fraunhofer Diffraction at a Double Slit

Consider a plane opaque shield containing two parallel slits each of width a separated by an opaque region of width d. The incident collimated light is of wavelength λ, and the slits are followed by a converging lens which focuses the light on a screen Z (Fig. 11.5).

At any point Q on the screen Z, the amplitude R_θ of the disturbance due to light emerging from slit 1 at an angle θ to the normal to the plane of the slit will be given by Equation 11.1:

$$R_\theta = \frac{R_0 \sin \alpha}{\alpha}$$

Superimposed on this disturbance will be that due to slit 2. These disturbances at Q from slits 1 and 2 will differ in phase because of the separation between the slits. To find this phase difference, consider that each slit is divided into equal parallel strips. For such strips central in the slits, it is seen from Fig. 11.5 that the path difference between rays emerging at the angle θ is $(d+a) \sin \theta$.

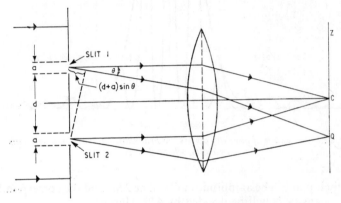

Fig. 11.5. Fraunhofer diffraction at a double slit

This same path difference applies to pairs of corresponding strips in the slits. Thus, for elemental strips at the bottom edges of the two slits, the path difference is again $(d+a) \sin \theta$, and also for such strips at the top edges. Indeed, any strip at a small distance s from the lower edge of one slit will have a companion at the same distance s above the lower edge of the other slit. The path difference between the disturbance at Q from the whole of slit 1 and that from slit 2 will hence also be $(d+a) \sin \theta$. The corresponding phase difference is $(2\pi/\lambda)(d+a) \sin \theta$.

From Equation 7.18, it follows that the resultant intensity at Q due to the contributions each of amplitude $(R_0 \sin \alpha)/\alpha$ from the two slits is given by:

$$I_\theta = 4R_\theta^2 \cos^2 \left[\frac{\pi(d+a) \sin \theta}{\lambda} \right] = \frac{4R_0^2 \sin^2 \alpha}{\alpha^2} \cos^2 \left[\frac{\pi(d+a) \sin \theta}{\lambda} \right]$$

The intensity at the centre C of the diffraction pattern will be that due to light emerging from the two slits along the normal

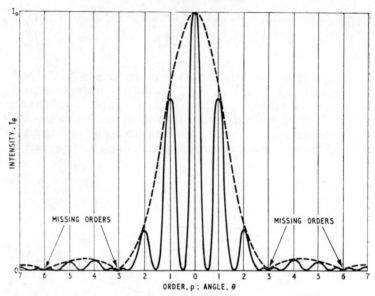

Fig. 11.6. Variation of intensity across the Fraunhofer diffraction pattern due to two slits, where $d = 2a$

to their plane. The amplitude at C will be $2R_0$, and the corresponding intensity I_0 will be decided by $4R_0^2$. Hence,

$$I_\theta = \frac{I_0 \sin^2 \alpha}{\alpha^2} \cos^2 \left[\frac{\pi(a+d) \sin \theta}{\lambda} \right] \qquad (11.8)$$

In Fig. 11.6, is plotted the intensity against θ for $d = 2a$. The dotted envelope of this curve is decided by the amplitude function $(I_0 \sin^2 \alpha)/\alpha^2$ and is characteristic of the variation of intensity due to a single slit. The actual more rapid variation of intensity shown by the full curve is characteristic of the cosine-squared fringes associated with line sources separated by a distance $a+d$.

This arrangement of two slits illuminated by normally incident collimated light of wavelength λ is that used by Young in his pioneer experiment on interference. In the mathematical account given in Section 8.2 of the intensity in the 'interference' pattern, the finite width of the slits was ignored. If in Equation 11.8 it is assumed that a is very small compared with d and also that θ is small, then

$$\alpha = \frac{\pi a \sin \theta}{\lambda} = \frac{\pi a \theta}{\lambda}$$

is very small; consequently $(\sin \alpha)/\alpha = 1$ and

$$I_\theta = I_0 \cos^2 \left(\frac{\pi d\theta}{\lambda}\right) \qquad (11.9)$$

This is the same as Equation 8.2 where d replaces l and $x/D = \theta$. The interference fringes are now of equal maximum intensity; the dotted envelope of Fig. 11.6 becomes a straight line parallel to the horizontal axis.

These considerations lead to a fuller understanding of diffraction as compared with interference. There is in reality no clear distinction between the two; both are the result of superposition of coherent waves. In the interference treatment of two slits, it is assumed that there are no separate zones within an individual slit which radiate light differing in phase. In the diffraction treatment, phase differences between separate zones in the wavefront at the slit are taken into account. The diffraction treatment is clearly necessary if the slit widths are significant.

In the diffraction pattern due to two slits, maxima will occur. These maxima will be of varying intensities, as shown by the envelope of the plot in Fig. 11.6. When this variation is small, the positions of the maxima will be approximately those of the 'interference' maxima; i.e. maxima will occur where

$$(a+d) \sin \theta = p\lambda \qquad (11.10)$$

p being an integer. Minima in the diffraction pattern will occur if in Equation 11.8 either $(\sin^2 \alpha)/\alpha$ is zero or $\cos^2 \{[\pi(a+d)\sin \theta]/\lambda\}$ is zero. The first will occur if $\alpha = p'\pi$, where p' is an integer; i.e. if

$$\alpha = \frac{\pi a \sin \theta}{\lambda} = p'\pi$$

or

$$\sin \theta = \frac{p'\lambda}{a} \qquad (11.11)$$

The second will occur if

$$\frac{\pi(a+d) \sin \theta}{\lambda} = (p''+0.5)\pi$$

where p'' is an integer; i.e. if

$$\sin \theta = \frac{(p''+0.5)\lambda}{a+d} \qquad (11.12)$$

Denoting $p = 1, 2, 3, \ldots$ as the successive orders where 'interference' maxima occur, *missing orders* will occur at locations where 'interference' maxima decided by Equation 11.10 are in the same place as 'diffraction' minima decided by Equation 11.11. Thus, whereas one would expect a maximum in the 'interference' fringes, this is obscured by a minimum due to 'diffraction', so the resultant intensity is zero or near zero. Missing orders will be decided by Equations 11.10 and 11.11, and will be where

$$\sin \theta = \frac{p\lambda}{a+d} = \frac{p'\lambda}{a}$$

i.e. where

$$\frac{p}{a+d} = \frac{p'}{a}$$

As p and p' are both integers, d must be an integral multiple of a. For example, if $d = 2a$, as in Fig. 11.6, missing orders p will occur where

$$\frac{p}{3a} = \frac{p'}{a}$$

i.e. at orders $p = 3$ for $p' = 1$, at $p = 6$ for $p' = 2$, and at $p = 9$, 12 and so on.

11.4 The Rayleigh Refractometer

This instrument is used to determine the refractive index of a gas or liquid; it rivals the Jamin interferometer (Section 10.5) but is more accurate. It can detect, by measurement of the change of refractive index, the presence in air of 0.01% hydrogen, 0.3% nitrogen, 0.006% helium, 0.03% carbon monoxide and 0.01% carbon dioxide. It can also be used, for example, to estimate the amount of sodium chloride or potassium chloride in solution in water with greater sensitivity than routine chemical analysis. For liquids, the refractive index can be measured to one unit in the seventh decimal place.

The Fraunhofer diffraction obtained with a pair of slits is utilised. Light from the slit S is collimated by the lens L_1 and then passes through slits S_1 and S_2, which are usually each 4 mm wide and 12 mm apart [Fig. 11.7(a)]. The separated parallel beams 1 and 2 emerging from the slits S_1 and S_2 traverse tubes T_1 and T_2 of equal lengths. Tubes T_1 and T_2 are closed at each end by parallel optically flat discs. For measurements on gases, these tubes are about 30 cm long and can be evacuated or filled with the gas at a pressure measured by an auxiliary manometer. For liquids, tubes T_1 and T_2 are plane parallel-sided cells usually of

1 cm length. On emerging from these tubes, the light beams pass through compensator plates C_1 and C_2, with plate C_1 in beam 1 and C_2 in beam 2. Together C_1 and C_2 form the compensator C, which is similar to that used in the Jamin interferometer (Section 10.5). The beams 1 and 2 are then received by an objective lens L_2, which converges them to overlap in the eyepiece E, where the two slit fringes are observed. With a separation of 12 mm between the slits S_1 and S_2, the cosine-squared interference fringes are very close together and considerable magnification by the eyepiece is essential for satisfactory observation.

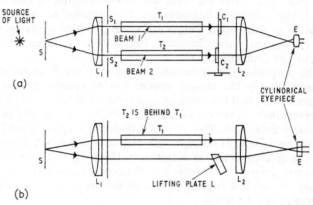

Fig. 11.7. The Rayleigh refractometer: (a) plan view; (b) elevation from side

In the measurement of the refractive index of a gas, one tube, say T_1, is evacuated and the other T_2 filled with the gas at a measured pressure. If the refractive index of the gas at this pressure is n, the increase in optical path of beam 2 over that of beam 1 is $(n-1)L$, where L is the length of either tube. The optical path therefore increases by $(n-1)L/\lambda$ wavelengths, where λ is the wavelength of the light from the monochromatic source. This number of interference fringes (cosine-squared fringes) will appear to traverse the cross-wires in the eyepiece, so n can be determined if this number is counted.

Referring to Equation 11.8 and the corresponding typical plot of Fig. 11.6, it is seen that the introduction of an optical path difference between the two beams does not alter α—and so does not alter the envelope of Fig. 11.6—because α is decided by the diffraction at any one aperture. However, the cosine-squared term will be affected because $[\pi(a+d)\sin\theta]/\lambda$ is decided by the path difference between disturbances from corresponding elements of the two slits S_1 and S_2, and this path difference is altered by $(n-1)L/\lambda$. In this sense, the instrument is an interferometer: the cosine-squared interference fringes are shifted, though the main variation in their intensities brought about by diffraction is unaffected.

It is usually impracticable to count monochromatic fringe shifts in the eyepiece. As in the Jamin interferometer, therefore, the compensator C has to be calibrated for light of a given wavelength. It is then necessary to find initially equality of optical paths when both tubes T_1 and T_2 are evacuated. This equality is determined by observation of white light fringes, i.e. the source is white light. After the fluid has been admitted to one of the tubes, the calibrated compensator C is rotated to restore path equality, so white light fringes are again obtained in the eyepiece, and the optical path difference is found from the measured rotation of the compensator C.

The magnification demanded of the eyepiece E to be able to observe clearly the closely spaced cosine-squared interference fringes is such that E is necessarily a cylindrical lens. This provides magnification in the only direction in which it is needed: perpendicular to the fringes. The use of a cylindrical eyepiece gives bright fringes because light is conserved. However, it renders impossible the use of cross-wires as a fiduciary mark against which fringe shifts are determined. Moreover, the cosine-squared interference fringes are so close together that a cross-wire is unsatisfactory for determining such shifts.

The Rayleigh refractometer is furnished, therefore, with an arrangement to provide a set of stationary reference fringes; the shift in the fringes used to measure optical path differences can be measured against these reference fringes with great precision. To provide the reference or fiduciary fringes, the light beams through the tubes T_1 and T_2 are from the top halves of the slit apertures S_1 and S_2 only. The light from the bottom halves of these slits does *not* pass through the tubes, so it gives a stationary double-slit fringe system in the focal plane of the objective lens L_2 [Fig. 11.7(b)]. To raise the lower stationary fringes (as observed by the cylindrical eyepiece E) to immediately below the upper fringes used for measurement, a 'lifting' plate L is used.

By the use of a lower fiduciary set of fringes against which shifts in the upper fringe pattern are measured, a path difference of 0·025 λ between the light beams 1 and 2 through tubes T_1 and T_2 can be measured. This leads to a method of determining the refractive index of a liquid, with T_1 and T_2 in the form of 1 cm length cells, which is about 100 times more accurate than a spectrometer determination.

A typical experiment for students with the Rayleigh refracto-meter is to verify the Gladstone and Dale law for a pure gas:

$$\frac{n-1}{\varrho} = \text{constant}$$

where n is the refractive index of the gas of density ϱ.

11.5 Michelson's Stellar Interferometer

A double slit in conjunction with a telescope can be used to determine the angular separation of two distant point sources of light (for example, the moons of Jupiter) or to measure the angular separation of the edges of a distant source of finite extent (for

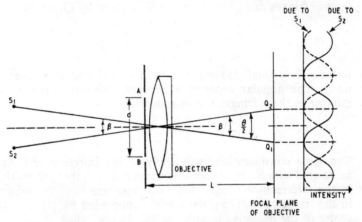

Fig. 11.8. Measurement of the angular diameter subtended by two distant strip sources of light by means of a telescope furnished with a double slit

example, the angular diameter of a star). The original idea was put forward by Fizeau in 1868 and used by Stefan in 1874. The eventual development of this method to enable very small angular separations to be determined is due to Michelson who, in 1920, introduced a 'periscope' arrangement of four mirrors to provide, in effect, two 'slits' with a separation of 20 ft or more.

Consider two parallel slits A and B, separated by a distance d and mounted immediately in front of the objective of a telescope (Fig. 11.8). The separation d is much larger than the width a of either slit. Let S_1 and S_2 be very thin distant strip sources of monochromatic light of wavelength λ. The angle subtended by

S_1S_2 at the centre of the telescope objective is β, and the distance from the slits to the focal plane of the objective is L, where $L >> d$.

The strip source of light S_1 will have an image at Q_1 in the focal plane of the telescope objective. The presence of the two slits will result in a fringe system about Q_1. Ignoring the slowly varying intensity in the pattern due to diffraction and considering only the interference fringes, Q_1 will be the position of the central maximum in these interference fringes due to S_1. Likewise, Q_2 will be the image in the focal plane of the strip source S_2 and Q_2 will be the position of the central maximum in the interference fringes due to S_2. The angle subtended by Q_1Q_2 at the centre of the telescope objective will be β, the angle subtended at this centre by S_1S_2.

The angular separation between adjacent maxima in the interference fringes due to strip source S_1 is given by Equation 11.10 as

$$\theta = \frac{\lambda}{d}$$

because θ is small (so $\sin \theta = \theta$), $a << d$, and p changes by unity. The angular separation between adjacent maxima and minima in these fringes is consequently:

$$\frac{\theta}{2} = \frac{\lambda}{2d}$$

The same considerations will apply to the interference fringes associated with S_2. If the slit separation d is such that the maxima in the interference fringes due to S_1 coincide with the minima in the fringes due to S_2, the angle β subtended by Q_1Q_2 at the centre of the objective is equal to $\theta/2$. Hence, when

$$\beta = \frac{\lambda}{2d} \tag{11.13}$$

there will be a uniform intensity of light in the focal plane of the telescope so the fringes will disappear.

The intensity in the focal plane, and hence in the field of view of the eyepiece, is uniform at this setting; this is seen from the fact that the intensity variation with angle θ measured from the centre of one of the patterns due to S_1, say, is given by Equation 11.8 to be of the form $I_0 \cos^2 (\pi d\theta/\lambda)$ because $(\sin^2 \alpha)/\alpha^2 = 1$, a is ignored compared with d, and θ is small so that $\sin \theta = \theta$. The intensity variation relative to the same centre in the pattern due to the other source S_2 is then $I_0 \cos^2 [\pi d(\theta+\beta)/\lambda]$, presuming that the two strip sources S_1 and S_2 are of the same intensity. When, from Equation 11.13,

$$\beta = \frac{\lambda}{2d}$$

the total intensity is given by

$$I_0 \left\{ \cos^2 \left(\frac{\pi \, d\theta}{\lambda} \right) + \cos^2 \left[\frac{\pi \, d(\theta + \lambda/2d)}{\lambda} \right] \right\}$$

$$= I_0 \left[\cos^2 \left(\frac{\pi \, d\theta}{\lambda} \right) + \cos^2 \left(\frac{\pi \, d\theta}{\lambda} + \frac{\pi}{2} \right) \right]$$

$$= I_0 \left[\cos^2 \left(\frac{\pi \, d\theta}{\lambda} \right) + \sin^2 \left(\frac{\pi \, d\theta}{\lambda} \right) \right]$$

$$= I_0$$

which is constant.

It is readily seen that Equation 11.13 can be extended to

$$\beta = \frac{(p+0 \cdot 5)\lambda}{d} \tag{11.14}$$

where p is an integer. At such positions, the field of view seen in the telescope eyepiece will be of uniform intensity. At other positions, interference fringes will appear.

The angle subtended by a single strip source of finite width extending over $S_1 S_2$ can also be determined. The Equation 11.13 will not apply, however, because the source is now considered as an infinite number of parallel elemental strips 1, 2, 3, ... extending from strip 1 at S_1 to strip n at S_2 [Fig. 11.9(a)]. Each of these elemental strip sources will result in an interference fringe system in which the intensity variations are of the cosine-squared form. These overlapping intensity curves will have central maxima at Q_1 corresponding to the strip 1 at S_1, and extending to Q_2 corresponding to strip n at S_2 [Fig. 11.9(b)]. The maxima and minima in the interference fringes will therefore be blurred. When Q_2 extends as far as the adjacent first maximum at Q_1' due to S_1, the 'blurring' will be such that the intensity in the focal plane of the telescope objective becomes uniform.

To establish the setting at which this intensity is uniform, any elemental strip within OS_1, the upper half of $S_1 S_2$, may be considered to have a companion strip at a distance of $S_1 S_2/2$ within OS_2, the lower half of $S_1 S_2$. These strips will produce complementary fringes resulting in uniform intensity when the angle they subtend at the telescope objective centre is given by Equation 11.13 to be $\lambda/2d$. The angle concerned will obviously be $\beta/2$,

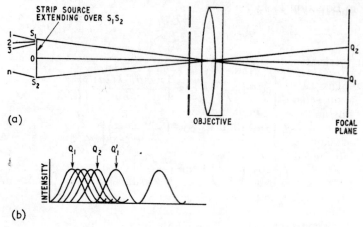

(a)

(b)

Fig. 11.9. Measurement of the angle subtended by a distant strip source of light of finite width

where β is the angle subtended by the full strip source S_1S_2 at the objective. Therefore, the condition for uniform intensity is

$$\beta = \frac{\lambda}{d} \qquad (11.15)$$

or, corresponding to Equation 11.14,

$$\beta = \frac{p\lambda}{d} \qquad (11.16)$$

where p is an integer.

If the single source is not a strip of finite width but a circular disc, more light will be contributed by its centre than by the periphery. Equation 11.15 then becomes:

$$\beta = \frac{1 \cdot 22\lambda}{d}$$

If $\lambda = 5{,}000$ Å $= 5 \times 10^{-5}$ cm and a star subtends an angle of $1''$ at the telescope, the required separation d between the pair of slits over the objective lens to give fringe disappearance (i.e. a uniformly illuminated field in the telescope eyepiece) will be given by the equation:

$$\frac{1}{3{,}600} \times \frac{2\pi}{360} = \frac{1 \cdot 22 \times 5 \times 10^{-5}}{d}$$

Therefore,

$$d = \frac{6 \cdot 1 \times 10^{-5} \times 12 \cdot 96 \times 10^{-5}}{2\pi} = 12 \cdot 6 \text{ cm}$$

However, many fixed stars subtend to an observer on Earth an angle of $0 \cdot 01''$ or less. To measure such an angle would demand a slit separation of $12 \cdot 6 \times 100$ cm or more, i.e. $12 \cdot 6$ m. As the slits must be accommodated within the objective of the telescope, this would demand impossibly large telescope lenses or mirrors of diameters up to 40 ft. It was for this reason that Michelson introduced his stellar interferometer.

Michelson's stellar interferometer avoids a prohibitively large telescope objective by the use of four mirrors M_1, M_2, M_3 and M_4 mounted on a long rigid girder in front of the objective; the outer mirrors M_1 and M_4, which extend well beyond the confines of the objective aperture, simulate the two slits [Fig. 11.10(a)].

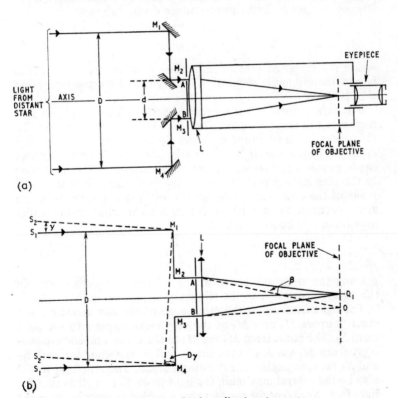

Fig. 11.10. *Michelson's stellar interferometer*

Light from the distant star is incident at an angle of 45° on to the outer mirrors M_1 and M_4. The reflected beams from M_1 and M_4 pass to mirrors M_2 and M_3 respectively and, from these, into the telescope objective lens L. Mirrors M_2 and M_3 are fixed at a separation d, which is equal to the separation between the slits A and B before the objective, and is less than the diameter of this telescope objective lens (or mirror); M_2 and M_3 are symmetrically disposed about the axis of the telescope. The outer mirrors M_1 and M_4 can be moved symmetrically to vary their separation D, where $D > d$.

In the field of view of the telescope eyepiece, focused on the focal plane of the telescope objective, appears an image of the star which is in the form of concentric diffraction rings crossed by a number of equally spaced linear interference fringes; these linear fringes are very close together, so an eyepiece of considerable magnifying power is needed to view them. The outer mirrors M_1 and M_4 are moved symmetrically until the linear interference fringes disappear. This disappearance occurs when

$$\gamma = \frac{1 \cdot 22\lambda}{D}$$

where γ is the angle subtended by the star at the telescope, but D is now the minimum separation of $M_1 M_4$ and *not* the separation d of the slits.

The arrangement of the mirrors M_1, M_2, M_3 and M_4 effectively increases the separation between the slits by the ratio D/d. To show that this is so, suppose S_1 and S_2 are the extreme edges of a diameter of the star [Fig. 11.10(b)], where S_1 is on the axis of the telescope and S_2 is slightly off this axis. The central maximum in the interference pattern caused by slits A and B at the focal plane of the objective due to light from S_1 is at Q_1, i.e. Q_1 is the zero-order interference fringe due to S_1. The adjacent first-order maximum will be at O, where

$$\beta = \frac{1 \cdot 22\lambda}{d} \qquad (11.17)$$

β being the angle subtended at the telescope objective between the adjacent maxima in the fringes.

The light from S_2, at the other edge of the star diameter, will reach mirrors M_1 and M_4 at the very small angle γ to the light from S_1. The paths from M_1 via M_2 to the slit A, and correspondingly from M_4 via M_3 to the slit B, will be the same for both S_1 and S_2 because angle γ is exceedingly small. Therefore, Q_1 would also be the central maximum for light from S_2 as well as that for light from S_1 *except for the fact that a path difference between the* *rays from S_1 and S_2 is introduced before the light reaches M_1 and M_4. This path difference, due to the angular separation of γ between S_1 and S_2, is $D\gamma$.* When

$$D\gamma = 1.22\,\lambda \tag{11.18}$$

the disturbance reaching Q_1 due to light from S_2 is 2π out of phase relative to the disturbance at Q_1 due to light from S_1. As the star extends from S_1S_2, comparison with the conditions leading to Fig. 11.9(b) shows that the linear interference fringes in the diffracted star image seen in the telescope eyepiece will disappear.

Comparison of Equations 11.17 and 11.18 shows that the presence of the mirrors M_1 and M_4 at a separation D, therefore, effectively increases the separation between the slits A and B in the ratio D/d. This increase is seen to be due to the ratio of the path difference that is introduced *before* the star light reaches the slits A and B to that which is introduced within the telescope *after* passage through the slits.

Star light is so weak that monochromatisation by a prism or grating is not possible, so a coarse filter passing a band of wavelengths is used. In an experiment conducted by Michelson with the 100 in diameter reflecting telescope at Mount Wilson, mirrors M_1, M_2, M_3 and M_4 were mounted on a rigid girder of total length 21 ft. Observations on the star Betelgeuse gave a minimum separation D between M_1 and M_4 at 307 cm (10 ft) for a first disappearance of linear interference fringes. The mean wavelength of the light passed by the filter used was 5.75×10^{-5} cm. The angular diameter γ of Betelgeuse was, therefore, found to be:

$$\gamma = \frac{1.22 \times 5.75 \times 10^{-5}}{307} = 2.28 \times 10^{-7}\ \text{rad} = 0.047''$$

The distance to Betelgeuse was found by parallax measurements; multiplication of this distance by its angular diameter gave the linear diameter to be 2.4×10^8 miles, which is 278 times the diameter of the Sun.

11.6 Fraunhofer Diffraction at Multiple Slits: The Diffraction Grating

An opaque shield contains N equally spaced parallel straight transparent slits each of width a, with neighbouring slits being separated by an opaque region of width b. Such an arrangement forms a *plane grating;* the integer N may be a small number but for practical diffraction gratings it is several thousand.

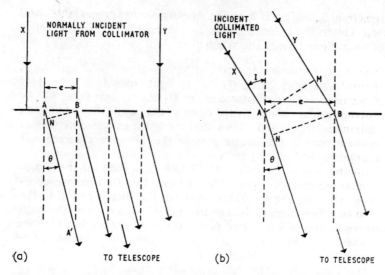

Fig. 11.11. The plane grating: (a) incident light is normal; (b) incident light at an angle I

Suppose the plane grating is illuminated by monochromatic light of wavelength λ from a collimator and that, after transmission through the grating, the light enters a telescope focused at infinity. The usual simple experimental arrangement therefore consists of the plane grating mounted on the rotatable table (the 'prism table') of a spectrometer. If the incident collimated light is incident normally on the plane grating [Fig. 11.11(a)], each of the transparent slits acts as a source to produce light in a direction at an angle θ to the normal and of amplitude R_θ given by Equation 11.1 as

$$R_\theta = \frac{R_0 \sin \alpha}{\alpha}$$

where $\alpha = (\pi a \sin \theta)/\lambda$, and R_0 is the amplitude in the direction where $\theta = 0$.

The phase of the light at the point A at the centre of any one slit is the same as that at the point B at the centre of the neighbouring slit and the same as that at the centres of all the other slits, provided that the incident light is normal. In the transmitted beam at an angle θ to the normal, the path difference between rays from A and B at the centres of neighbouring slits is AN, where BN is drawn perpendicular to AA' in Fig. 11.11(a). Therefore,

$$BN = e \sin \theta$$

where $e = a+b$, which is the separation between neighbouring slit centres. If

$$e \sin \theta = p\lambda \tag{11.19}$$

where p is an integer, the path difference between light from successive slits is an integral number of wavelengths; hence, reinforcement due to constructive interference gives the *principal maxima* in directions θ decided by Equation 11.19. The central maximum occurs at $\theta = 0$ when $p = 0$; the next principal maximum is when $p = 1$ and is the first-order maximum, followed by second, third and, in general, kth order maxima at values in Equation 11.19 corresponding to $p = 2$, 3 and, in general, k.

If the collimated light is incident at an angle I to the normal [Fig. 11.11(b)], AM—drawn perpendicular from A to BY at M— will be a wavefront in the incident beam. The path difference between transmitted rays leaving A and B at an angle θ to the normal is then:

$$AN - BM = e(\sin \theta - \sin I)$$

The equation deciding the various angles θ at which principal maxima occur is then

$$e(\sin \theta - \sin I) = p\lambda \tag{11.20}$$

where p is an integer.

With incident white light corresponding to various values of λ extending continuously over the range from approximately 4,000 Å to 7,000 Å, the light transmitted through the grating will give rise to a continuous spectrum in the first order, a second spectrum in the second order, a third spectrum in the third, and so on. For spectroscopy, a plane grating is made by ruling on glass with a diamond attached to a dividing engine a large number of fine parallel lines (say 5,000 to the centimetre). These scratches are nearly opaque and are separated by clear glass (the 'slits'). For cheapness, replicas in plastic are made from the master grating on glass. If there are N lines per centimetre the separation between neighbouring slit centres $e = 1/N$. With normal incidence, Equation 11.19 can then be written:

$$\frac{\sin \theta}{N} = p\lambda$$

With $N = 5,000$ per centimetre,

$$\theta = \sin^{-1}(p\lambda N)$$
$$= \sin^{-1}(5,000\,\lambda) \text{ in the first order}$$
$$= \sin^{-1}(10,000\,\lambda) \text{ in the second order}$$

In the red of the spectrum, at $\lambda = 7{,}000$ Å $= 7 \times 10^{-5}$ cm, for the first order

$$\theta = \sin^{-1}(5 \times 10^3 \times 7 \times 10^{-5}) = \sin^{-1} 0.35 = 20°30'$$

and for the second order

$$\theta = \sin^{-1}(10^4 \times 7 \times 10^{-5}) = \sin^{-1} 0.7 = 44°24'$$

In the violet end of the spectrum, at $\lambda = 4{,}000$ Å $= 4 \times 10^{-5}$ cm, for the first order

$$\theta = \sin^{-1}(5 \times 10^3 \times 4 \times 10^{-5}) = \sin^{-1} 0.2 = 11°32'$$

and for the second order

$$\theta = \sin^{-1}(10^4 \times 4 \times 10^{-5}) = \sin^{-1} 0.4 = 23°35'$$

Note that in the second-order spectrum the angle θ for violet light is $23°35'$, which is larger than that for the extreme red light in the first-order spectrum, so there will be some overlap between the first and second order spectra.

So far, in the treatment of Fraunhofer diffraction by N equal parallel slits separated by equal opaque spaces, only the positions of the principal maxima have been found. The variation in intensity through the diffraction pattern due to N parallel, equal slits is now considered. Beginning at one edge of the grating, let the slits be numbered successively $1, 2, 3, \ldots, N$. These slits are each of width a, and the separation between centres of neighbouring slits is e. Due to slit 1 only, the amplitude of the disturbance in the focal plane of the objective of the viewing telescope is given by Equation 11.1 as

$$R_\theta = \frac{R_0 \sin \alpha}{\alpha}$$

where the telescope axis is set at angle θ to the normal to the grating, $\alpha = (\pi a \sin \theta)/\lambda$, and $R_\theta = R_0$ when $\theta = 0$.

The amplitude due to all the slits is the same. However, slit 2 will produce a disturbance differing in phase by δ from that due to slit 1, and this same phase angle δ will exist between disturbances from any two neighbouring slits. The phase angle δ is a consequence of the path difference $e(\sin \theta - \sin I)$ considered previously, where I is the angle of incidence of the incident collimated light. Whereas R_θ represents the contribution to the resultant amplitude due to slit 1, that due to slit 2 will be $R_\theta \cos \delta$, that due to slit 3 will be $R_\theta \cos 2\delta$ and, in general, that due to the Nth slit will be $R_\theta \cos(N-1)\delta$. Therefore, the resultant disturbance due to N slits in the direction θ is given by the real part of the sum

$$R_\theta[1 + \exp j\delta + \exp 2j\delta + \ldots + \exp(N-1)j\delta]$$

where $\delta = 2\pi e(\sin\theta - \sin I)/\lambda$ because it is the phase difference corresponding to a path difference of $e(\sin\theta - \sin I)$. This series is a geometrical progression of which the sum is $R_\theta(1 - \exp Nj\delta)/(1 - \exp j\delta)$.

The amplitude of the resultant $R_{N\theta}$ (i.e. the amplitude of the resultant disturbance due to N slits in the focal plane of the telescope objective in the direction θ) is the modulus of this expression. The intensity is proportional to the square of the amplitude, which is given by multiplying this expression by its complex conjugate $R_\theta[1 - \exp(-Nj\delta)]/[1 - \exp(-j\delta)]$. Therefore,

$$R^2_{N\theta} = R^2_\theta \frac{(1 - \exp Nj\delta)[1 - \exp(-Nj\delta)]}{(1 - \exp j\delta)[1 - \exp(-j\delta)]}$$

$$= R^2_\theta \frac{2 - [\exp Nj\delta + \exp(-Nj\delta)]}{2 - [\exp j\delta + \exp(-j\delta)]}$$

$$= R^2_\theta \frac{1 - \cos N\delta}{1 - \cos \delta}$$

$$= R^2_\theta \left[\frac{\sin(N\delta/2)}{\sin(\delta/2)}\right]^2$$

Putting $R_\theta = (R_0 \sin\alpha)/\alpha$ and $\beta = \delta/2 = \pi e(\sin\theta - \sin I)/\lambda$ gives

$$I_{N\theta} = I_0 \left(\frac{\sin\alpha}{\alpha}\right)^2 \left(\frac{\sin N\beta}{\sin\beta}\right)^2 \qquad (11.21)$$

where $I_{N\theta}$ is the intensity in the direction θ due to N slits, and I_0 is the intensity at the centre of the diffraction pattern where $\theta = 0$. Further,

$$R_{N\theta} = R_0 \frac{\sin\alpha}{\alpha} \frac{\sin N\beta}{\sin\beta} \qquad (11.22)$$

gives $R_{N\theta}$, the amplitude in the direction θ due to N slits, where R_0 is the amplitude at $\theta = 0$.

The principal maxima in the intensity distribution occur at $e\sin\theta = p\lambda$ (Equation 11.19) when the incident light is normal. At such principal maxima

$$\beta = \frac{\pi e \sin\theta}{\lambda} = p\pi$$

where p is an integer. The numerator and denominator of Equation 11.22 are then both zero because $\sin p\pi = 0$. The expression for the amplitudes R_{max} at the principal maxima is, therefore,

$$R_{max} = R_0 \frac{\sin \alpha}{\alpha} \lim_{\beta \to p\pi} \frac{\sin N\beta}{\sin \beta}$$

$$= R_0 \frac{\sin \alpha}{\alpha} \lim_{\beta \to 0} \frac{\sin N\beta}{\sin \beta}$$

$$= R_0 \frac{\sin \alpha}{\alpha} \lim_{\beta \to 0} \left(\frac{N \sin N\beta}{N\beta} \frac{\beta}{\sin \beta} \right)$$

$$= NR_0 \frac{\sin \alpha}{\alpha} \tag{11.23}$$

and the corresponding intensities at the principal maxima are given by

$$I_{max} = N^2 I_0 \frac{\sin^2 \alpha}{\alpha^2} \tag{11.24}$$

which is N^2 times that due to one slit alone.

In Equation 11.21, $[(\sin \alpha)/\alpha]^2$ is the diffraction term—decided by the finite width of the slit—whereas $[(\sin N\beta)/\sin \beta]^2$ is the interference term decided by the N slit sources with a progressive phase difference on going from one slit to the others.

In the interference term, minima of zero intensity, corresponding to dark fringes, occur when $\sin N\beta = 0$ but $\sin \beta$ is finite. This happens when $N\beta = \pi, 2\pi, 3\pi$ and, in general, $m\pi$ where m is an integer (except that m cannot be zero or an integral multiple of N because then $\beta = m\pi/N = p\pi$, where p is an integer, corresponding to the positions of the principal maxima). Between neighbouring principal maxima, separated in angle by $\beta = \pi$, β will equal π/N to give minima at $N-1$ positions. There are thus $N-1$ minima between neighbouring maxima. Secondary maxima, of considerably lower intensity than the principal maxima, will occur between neighbouring minima; there will consequently be $N-2$ secondary maxima.

It is instructive to work out fully a specific example. Let the separation e between neighbouring parallel slit centres in a grating be 0·05 cm and the width a of each slit be 0·025 cm. Suppose there are $N = 6$ such slits in the grating, and that the wavelength λ of the incident light is 5,000 Å $= 5 \times 10^{-5}$ cm. The diffraction term in Equation 11.22 for the resultant amplitude becomes, on normalising to $R_0 = 1$,

$$\frac{\sin \alpha}{\alpha} = \frac{\sin [(\pi a \sin \theta)/\lambda]}{(\pi a \sin \theta)/\lambda} = \frac{\sin[(\pi \times 0·025 \times \sin \theta)/(5 \times 10^{-5})]}{(\pi \times 0·025 \times \sin \theta)/(5 \times 10^{-5})}$$

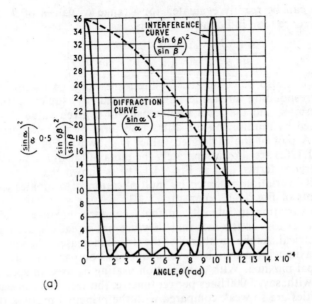

(a)

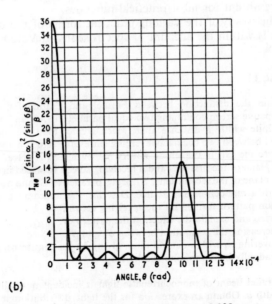

(b)

Fig. 11.12. Fraunhofer diffraction at six slits where a = 0·025 cm, e = 0·05 cm and λ = 5×10⁻⁵ cm: (a) the variation with angle θ of the diffraction term and the variation of the interference term; (b) the variation of intensity with angle θ

which can be readily evaluated for a range of values of θ. The interference term in Equation 11.22 is

$$\frac{\sin N\beta}{\sin \beta} = \frac{\sin[(6\pi \times 0\cdot05 \times \sin \theta)/\lambda]}{\sin[(\pi \times 0\cdot05 \times \sin \theta)/\lambda]}$$

which can also be computed for the same range of values of θ. Corresponding intensities are obtained by squaring these expressions, and the resultant intensity at a given value of θ is obtained by multiplying the diffraction term by the interference term. A plot of the calculated values is shown in Fig. 11.12; in Fig. 11.12(a) are shown separately the plots of the diffraction and interference terms, and in Fig. 11.12(b) is shown the combined intensity variation obtained by multiplication of ordinates from the plots of Fig. 11.12(a).

The variation in intensity in the secondary maxima is noteworthy, as also is the fact that they are not equally spaced and symmetrical. As the number N of the slits is increased, the secondary maxima intensities become smaller in comparison with the principal maxima. With a diffraction grating as used in spectroscopy with, say, 5,000 lines per centimetre, the secondary maxima intensities are so weak compared with the principal maxima that they are absent for most practical purposes.

Further study of the various practical forms of the diffraction grating is within the account of spectroscopy in Volume 2.

Exercise 11

1. Explain the distinction between Fresnel and Fraunhofer diffraction. An opaque screen containing two equal parallel slits, S_1 and S_2 (of small but finite width, a distance apart much greater than the slit width) is placed between two coaxial converging lenses, the screen being perpendicular to the axis of the lenses. A very narrow slit S_3 is placed in the first focal plane of the first lens and a receiving screen is placed in the second focal plane of the second lens, S_3 being parallel to S_1 and S_2. Behind S_3 there is a monochromatic source. Describe the appearance of the diffraction pattern formed on the receiving screen.
Describe and explain the effect on the pattern of:
(a) increasing the separation between S_1 and S_2,
(b) covering S_1 with a neutral filter, giving 50 per cent transmission and
(c) gradually increasing the width of S_3. (L. P.)

2. A parallel beam of monochromatic light is incident normally on a slit of width a. Obtain an expression for the light distribution in the focal plane of a converging lens placed after the slit. How is the distribution altered if light over one half of the slit is retarded by $\lambda/2$ with respect to the other half? (L. P.)

3. Derive an expression for the intensity distribution in the Fraunhofer diffraction pattern given by a slit.
Calculate, for normally incident monochromatic light, the ratio of the intensities of the maxima of the first and second orders of diffraction.
(L. P.)

4. Distinguish between Fresnel and Fraunhofer diffraction. Obtain an expression for the variation of intensity in the Fraunhofer diffraction pattern from a single slit. (L. G.)

5. Parallel monochromatic light is incident normally on a slit, passes through a converging lens near the slit, and is brought to a focus on a screen. Obtain an expression for the intensity distribution on the screen. Point out (a) the inclinations to the axis of the system for which minima occur (b) the conditions under which maxima occur.
If the incident light is white and slit has a width of 0·080 cm and the lens has a focal length of 80 cm find the wavelengths in the visible region which are missing on the screen at a distance of 0·30 cm from the axis of the system. (L. G.)

6. An opaque screen has two parallel slits cut in it with a separation of 0·020 cm between their centres, and is placed on the table of a spectrometer so that light of wavelength 5.46×10^{-5} cm falls normally from the collimator on it. Describe and explain what is seen in the spectrometer telescope. How would the observed pattern change if the source were replaced by (a) another monochromatic source of longer wavelength, and (b) by a white light source? (L. Anc.)

7. Briefly distinguish between Fraunhofer and Fresnel diffraction. Monochromatic light emerging from a collimator falls normally on a diaphragm containing two identical slits parallel to the collimator slit and placed in front of a converging lens. Deduce an expression for the distribution of light in the focal plane of this lens and discuss the effects observed when the width of the collimator slit is increased. Describe briefly the application of this arrangement to the measurement of stellar diameters (L. P.)

8. A vertical screen in which two narrow vertical slits are cut with a separation of 0·0059 cm between their centres is placed on the table of a spectrometer so that light of wavelength 5.89×10^{-5} cm falls normally on it. Describe and explain what is seen in the telescope.
What changes would be observed if (a) the sodium source were replaced by a white light source, and (b) if the two slits were replaced by eight equally spaced slits the source of light being in turn either (1) a sodium source and (2) a white light source? (L. Anc.)

9. Describe the use of the Rayleigh refractometer for the accurate determination of the refractive index of a gas.
Estimate the smallest molecular fraction of helium in a specimen of air at 400°C and 100 mmHg pressure, detectable in a tube of length 10 cm. Assume that a fringe shift of 1/25th of the spacing between fringes is

detectable and that for sodium light the refractive indices at S.T.P. are $(\mu_{air} - 1) = 2.92 \times 10^{-4}$, $(\mu_{He} - 1) = 0.35 \times 10^{-4}$, $\lambda_D = 5893$ Å. (L. P.)

10. Describe the design and method of use of either a Rayleigh or a Jamin refractometer for the determination of the refractive index of a gas.
    A determination of the refractive index of air, for the D-line at a pressure of 2 atmospheres and temperature 30°C, has given the result 1·000527. Deduce the refractive index at S.T.P. (L. P.)

11. Describe in detail an accurate method for the determination of the refractivity of a gas.
    In what problems in physics is an accurate knowledge of the refractive index of air important? (L. P.)

12. Describe and explain an interferometer suitable for measuring the refractivity $(n - 1)$ of air under standard conditions of temperature and pressure. Show how the conditions necessary for interference to take place are satisfied in the instrument described.
    A sample of helium is suspected of being contaminated with air. What is the smallest percentage contamination detectable with an interference refractometer having tubes of length 30 cm and in which a displacement equal to one-twentieth of a fringe spacing is just observable? For a wavelength of 5893 Å the refractivity of helium is 3.5×10^{-6}, and that of air is 2.92×10^{-4}, both at S.T.P. (L. G.)

13. Describe the optical features of an interferometer useful for determining the refractive index of a gas, and give the theory of the formation of the fringes.
    The refractive index of air at S.T.P. is 1·000292. Making some reasonable assumption about the smallest detectable fringe shift, estimate the least fractional change in pressure that may be detected by interference in a tube 20 cm long, when using light of wavelength 5000 Å. (L. G.)

14. Write an essay on the determination of the refractive indices of solids, liquids and gases. (L. Anc.)

15. Describe and explain the action of an instrument for determining the angular diameter of a star. Estimate numerically the smallest angular diameter resolvable in the visible region if a girder 10 metres in length is available. (L. P.)

16. Describe a two-beam interference method for the determination of the refractive index of a gas.
    In an experiment using such a method the apparatus was maintained at 20°C and a slow leakage of air (refractive index at S.T.P. = 1·00029) into a tube of length 10 cm resulted in one fringe per minute (at wavelength 5800 Å) passing a point in the field of view. Calculate the rate of change of the pressure of the air in the tube. (L. P.)

17. Three parallel slits, each of width b, are cut in a screen so as to be separated by opaque strips also of width b, and the screen is illuminated by monochromatic parallel light falling normally on it. The transmitted light is observed in a telescope. Derive expressions for the positions of zero intensity in the respective diffraction patterns when the light is allowed to pass through (a) one slit, (b) two adjacent slits and (c) all three slits. Make a rough sketch of the diffraction pattern in each case.   (L. P.)

18. A beam of parallel monochromatic radiation of wavelength 5×10^{-5} cm falls normally on a screen in which are cut six parallel apertures each of width 0·01 cm, corresponding points of adjacent slits being at a distance 0·04 cm apart. Draw a graph approximately to scale showing the distribution of light in the focal plane of a converging lens of focal length 100 cm which collects all the transmitted light. Assuming that the angular distribution of light intensity due to Fraunhofer diffraction by a single slit is given by the expression

$$I = \frac{A^2 \sin^2 \alpha}{\alpha^2}$$

derive any formulae you use.   (L. P.)

19. Derive, with full explanation, an expression for the intensity of light diffracted in the nth order by a plane grating consisting of parallel opaque strips of width a separated by transparent spaces of width b, when light of intensity I and wavelength λ falls normally on the grating. (Assume a and b large compared with λ.)

The following angles of diffraction and estimates of intensity were obtained with such a grating used at normal incidence with light of wavelength 5000 Å.

0°17′	Very strong	1°43′	Weak
0°34′	Strong	2°18′	Weak
0°51′	Weak	2°35′	Weak
1°9′	Weak	3°9′	Weak
1°26′	Medium		

Obtain from this data such information as you can about the values of a and b.   (L. P.)

20. A reflecting astronomical telescope focuses light from a star on to a photographic plate using a concave mirror of diameter 120 cm and focal length 700 cm. Describe and explain the photographs which may be obtained when the front of the tube is covered with (a) no screen, (b) an opaque cover with two circular apertures 12 cm in diameter and 100 cm apart, (c) a transmission grating with a grating space of 1 mm, (d) a square-mesh wire net with wires 1 mm thick spaced apart by 2 cm and (e) a mosaic screen complementary to the wire mesh in (d).   (L. P.)

21. Give the simple theory of the plane diffraction grating consisting of a large number of parallel slits each of width a separated by opaque bars of width b. Given the wavelength of the mercury green line, describe and

explain how a plane diffraction grating may be used to determine the difference between the wavelengths of the D_1 and D_2 lines of sodium.

(L. G.)

22. A parallel beam of monochromatic light of wavelength λ is incident obliquely on a plane transmission grating consisting of parallel transparent rulings of width b, the diffracted light is focussed by a converging lens. Obtain the equation which gives the positions of the principal maxima in the focal plane of the lens. Explain with the necessary theory what will be observed if $a = 2b$.

Explain briefly a method of testing the uniformity of the rulings of a plane grating. (L. G.)

23. Explain the simple theory of the plane diffraction grating using either transmission or reflection. What is the greatest number of orders that could be observed, using a plane grating with 3,000 lines per cm, and normally incident light of wavelength 5460 Å? If light of all wavelengths from 4000 Å to 7000 Å were used what wavelengths would be superposed on 5460 Å in the highest of these orders? (L. Anc.)

24. Explain the production of a spectrum by a plane transmission grating. A grating has 6000 lines per cm width, and a beam of parallel light falls on it at 45° to normal. Calculate the visible wavelength that should be present in the transmitted beam diffracted in the normal direction.

(L. Anc.)

25. Give the simple theory of the plane diffraction grating, and describe its use for comparing the wavelength of two spectrum lines. When monochromatic light is incident normally on a plane diffraction grating, the first order spectra for the line are found at 15° on either side of the central image. Find the positions of the first order spectra when the angle of incidence is 20°. (L. Anc.)

26. Give the theory of the plane transmission grating and obtain an expression for its resolving power. How would the phenomenon of overlapping spectra of different orders be affected by immersing the grating in a dispersive medium? (L. P.)

27. A single narrow slit is illuminated by monochromatic light. Obtain the approximate directions of maxima and minima.

The slit is now covered with a partially transmitting silvered glass slide with an extremely fine line ruled down the centre. Assuming the light transmitted through the fine line is of greater amplitude than that transmitted by the rest of the slit in all but the forward direction, discuss what changes in the intensity pattern as a function of angle you would expect.

(M. P.)

CHAPTER 12

Resolving Power

The term 'resolving power' is employed in two connections: to specify the extent to which an optical instrument, such as a telescope or microscope, is able to reveal detail of the structure of the object viewed; and to specify the extent to which a spectroscopie instrument is able to reveal the structure of the spectrum viewed. The first usage concerns the ability of the instrument to provide in its field of view separate distinct images of two very small objects which are close together. The smaller this object separation can be, the smaller is the *limit of resolution* of the instrument. For a microscope the smallest *linear* separation between 'points' or 'lines' in the object is specified; for a telescope, the specification is of the smallest *angular* separation. The second concern is with the ability of a spectroscope to distinguish a small separation of wavelength between two lines of wavelengths λ and $\lambda + \delta\lambda$ in a spectrum. The appropriate term is *chromatic resolving power*, specified as $\lambda/\delta\lambda$. The better the instrument in this connection, the larger its chromatic resolving power.

12.1 Limit of Resolution

The limit of resolution of an optical instrument, such as a microscope or telescope, will be decided by diffraction. An instrument having an objective lens of circular aperture will provide an image of a point object, the image being in the form of a series of concentric alternately dark and bright rings around the central maximum called the *Airy disc*. This bright central disc is centred on the image point. Neighbouring point objects will give rise to neighbouring overlapping sets of diffraction rings around the neighbouring 'point' images in corresponding Airy discs. A criterion is needed to specify when these overlapping diffraction patterns can be said to be distinguishable from one another, so the Airy disc images can be seen to be separate. The *Rayleigh criterion* (Section 12.2) is frequently used in this connection; alternative criteria have also been suggested.

The limit of resolution of an optical instrument will be adversely affected by aberrations of the lenses and/or mirrors from which it is constructed. Even an instrument completely free from such aberrations will, however, have a finite limit of resolution because of diffraction brought about by the inevitable apertures involved. For example, the normal human eye has a limit of resolution of 1' approx., corresponding to a linear separation of $\frac{1}{15}$ mm at the least distance of distinct vision of 25 cm. At a distance of 50 m, the normal eye would be able to discern a linear separation between points of about $(5 \times 10^3 \times 2\pi)/(60 \times 360) = 1.45$ cm. The defective eye, unaided by spectacles, would not be capable of this resolution because of aberrations brought about by astigmatism or incorrect separation between pupil and retina.

12.2 The Rayleigh Criterion

Curve 1 of Fig. 12.1(a) shows the variation in intensity of the diffraction pattern around an image Q_1; in curve 2 is shown the variation around a separated image Q_2 in the same field of view,

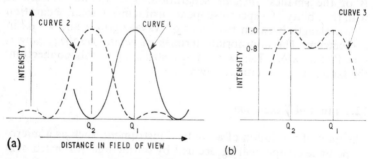

Fig. 12.1. The Rayleigh criterion

where the central maximum intensities at Q_1 and Q_2 are equal and normalised to unity. According to the Rayleigh criterion, *two diffraction patterns with equal intensities may be said to be resolved when the central maximum of one pattern is not nearer than the position of the centre of the first minimum of the neighbouring pattern.* At the limit of resolution, therefore, the central maximum of curve 1 is in the same position as the first minimum in curve 2. Then, the resultant curve 3—shown dotted in Fig. 12.1(b)—will exhibit a diminution in intensity midway between the central maxima of curves 1 and 2.

If the limiting aperture through which the image-forming light passed was in the form of a slit of width a, the first minimum in one of the diffraction patterns would be separated by an angle $\theta = \lambda/a$ (Equation 11.7) from the central maximum. Midway between the central maxima in the two diffraction patterns around Q_1 and Q_2 will, at the limit of resolution, correspond to an angle of $\theta/2 = \lambda/2a$. At this angle, the intensity in one image alone will be

$$I_\theta = I_0 \frac{\sin^2 \alpha}{\alpha^2} = \frac{\sin^2 \alpha}{\alpha^2}$$

from Equation 11.2, where $I_0 = 1$, and

$$\alpha = \frac{\pi a \sin \theta}{\lambda} = \frac{\pi a \theta}{\lambda} = \frac{\pi a(\lambda/2a)}{\lambda} = \frac{\pi}{2}$$

Therefore,

$$I_\theta = \frac{\sin^2 (\pi/2)}{(\pi/2)^2} = \frac{4}{\pi^2} \tag{12.1}$$

The intensity at the centre of the saddle in Fig. 12.1(b) will be double this value because there are two images of equal intensity. Hence, the intensity at the saddle centre is $8/\pi^2$, i.e. 0·8 approx. of that at the central maxima on either side. The Rayleigh criterion, when concerned with light traversing a slit aperture, therefore corresponds to the ability to distinguish two images as separated if the intensity of light in the field of view midway between their centres is not greater than 80% of the maximum intensity prevailing at the centre of the diffraction patterns of either of the two images.

In the more usual case where the limiting aperture in the optical instrument is circular, this result is modified. In Fig. 12.2, let AB represent the circular exit pupil of an optical instrument constructed from spherical lenses and/or mirrors. The axis of this instrument is XX'. When the instrument is focused on a point object on XX', converging spherical waves in the image space will leave AB to form the image Q on XX'. This centre Q will be surrounded by concentric diffraction rings. The first minimum occurs in this

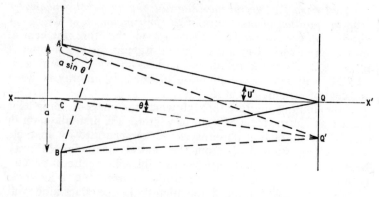

Fig. 12.2. The resolution of an optical system of spherical lenses or mirrors

diffraction pattern at Q'. If λ is the wavelength of the light forming the image, it can be shown that the position of Q' is such that the optical paths from Q' to the extreme edges of the exit pupil differ by $1 \cdot 22\lambda$, i.e.

$$Q'A - Q'B = 1 \cdot 22\lambda$$

Therefore,

$$a \sin \theta = 1 \cdot 22\lambda \qquad (12.2)$$

where a is the diameter of the exit pupil and θ is the direction of the minimum at Q' (i.e. the angle CQ' makes with XX', C being the centre of the exit pupil).

If Q is in the focal plane of an objective lens or mirror of focal length f, there is consequently a dark ring centred on Q of radius QQ' given by:

$$QQ' = f\theta = \frac{1 \cdot 22 f\lambda}{a} \qquad (12.3)$$

Now,

$$\frac{a/2}{f} = \sin U'$$

where U' is the semi-angle of the cone of light in the image space emerging from the exit pupil. Therefore the radius QQ', known as the radius of the Airy disc at the centre of the concentric diffraction rings surrounding the point image, is given by:

$$\text{radius of Airy disc} = \frac{1 \cdot 22 f\lambda}{a} = \frac{1 \cdot 22\lambda a}{2a \sin U'} = \frac{0 \cdot 61\lambda}{\sin U'}$$

If λ is the wavelength of the light measured in a vacuum whereas the refractive index of the image space is n', the wavelength concerned is then λ/n'; hence,

$$\text{radius of Airy disc} = \frac{0\cdot 61\lambda}{n' \sin\ U'} \qquad (12.4)$$

This equation decides the angular separation between two images of equal intensity in the focal plane of an objective which can be regarded as distinct in accordance with the Rayleigh criterion. The centre of the Airy disc associated with one image will then coincide with the first dark ring in the diffraction pattern due to the second image.

12.3 The Limit of Resolution of a Telescope Objective

In Fig. 12.3, the objective lens L of a telescope produces in its focal plane images Q_1 and Q_2 respectively of distant objects P_1 and P_2. The limit of resolution at which Q_1 and Q_2 are just

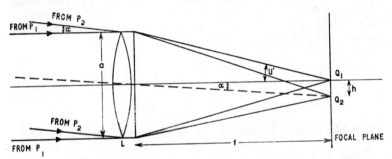

Fig. 12.3. The resolution of a telescope objective

distinguishable is when their linear separation h is equal to the radius of the Airy disc; so, from Equation 12.4,

$$h = \frac{0\cdot 61\lambda}{\sin\ U'}$$

where λ is the wavelength of the light from the objects, presumed to be sources of equal intensity, $n' = 1$, and U' is the semi-angle of the cone of image-forming rays leaving the objective.

If α is the angle subtended at the objective by objects P_1 and P_2 which are just resolvable as giving separate images,

$$\alpha = \frac{h}{f}$$

where f is the focal length of the telescope objective. Therefore,

$$\alpha = \frac{0 \cdot 61\lambda}{f \sin U'}$$

But

$$f \sin U' = \frac{a}{2}$$

where a is the diameter of the objective. Therefore,

$$\alpha = \frac{0 \cdot 61\lambda}{a/2} = \frac{1 \cdot 22\lambda}{a} \qquad (12.5)$$

To obtain a very small angle of resolution at the limit, the diameter of the objective must clearly be very large. Owing to the difficulty of making very large corrected objective lenses, astronomical telescopes providing extremely high resolution utilise a paraboloidal concave mirror objective. The largest of these is at Mount Palomar in California, which has a mirror of diameter 200 in (500 cm). For light of wavelength 5×10^{-5} cm, this provides a limit of resolution of angle α given by:

$$\alpha = \frac{1 \cdot 22 \times 5 \times 10^{-5}}{500} = 1 \cdot 22 \times 10^{-7} \text{ rad}$$

12.4 The Limit of Resolution of a Microscope Objective

In Fig. 12.4, L is the objective of a microscope which is of short focal length. The object viewed is placed at, or just inside, the focal plane of the objective, so the semi-angle of the cone of

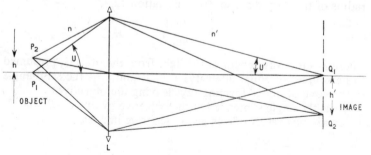

Fig. 12.4. The resolution of a microscope objective

rays diverging from the object point to the extremes of the aperture of the objective is large. Let P_1 be an axial object point, and P_2 a slightly non-axial object point separated by h from P_1 and in the same plane perpendicular to the microscope axis. The image of P_1 is at Q_1, and that of P_2 is at Q_2. Assuming that P_1 and P_2 are independently radiating point sources of light of equal intensity, both centres Q_1 and Q_2 will be surrounded by first dark rings of radius equal to that of the Airy disc; this radius is given by Equation 12.4 as $0.61\ \lambda/(n' \sin U')$, where λ is the wavelength of the light *in vacuo* from the objects, n' is the refractive index of the image space, and U' is the semi-angle of the image-forming rays leaving the objective. The separation h' between the image centres at Q_1 and Q_2 at the limit of resolution will be when the central maximum at Q_2 falls on the first dark ring around Q_2. This occurs when

$$h' = \frac{0.61\lambda}{n' \sin U'} \qquad (12.6)$$

The objective L has to be capable of providing a lateral magnification which is the same for all its zones, so the Abbe sine condition must be obeyed; i.e. in accordance with Equation 2.2,

$$m = \frac{h'}{h} = \frac{n \sin \theta}{n' \sin \theta'} = \text{constant}$$

where n is the refractive index of the object space. In particular, for the cone of rays from the object to the extremes of the objective which have a semi-angle of U:

$$m = \frac{h'}{h} = \frac{n \sin U}{n' \sin U'} = \text{constant}$$

Therefore,

$$h = \frac{n'h' \sin U'}{n \sin U}$$

Substituting for h' from Equation 12.6 gives:

$$h = \frac{0.61\lambda}{n \sin U} \qquad (12.7)$$

This equation gives the least separation h between object points which can give rise to images distinguishable as separate in the field of view observed by the microscope eyepiece.

The quantity $n \sin U$ is known as the *numerical aperture* (N.A.) of the microscope objective. Clearly $\sin U$ cannot exceed unity, though high-class microscope objectives are made for which $\sin U$

is 0·9. To increase n, an oil-immersion objective may be used in which the space between the microscope slide and the objective is filled with a suitable oil. The maximum numerical aperture achievable with an immersion objective utilising an oil of refractive index 1·5 is $0·9 \times 1·5 = 1·35$. The use of higher refractive index oils has led to the design of immersion objectives with a numerical aperture as much as 1·6.

In the derivation of Equation 12.7, it is assumed that the object viewed consists of separated point sources of light which radiate independently. In practice, a microscope slide illuminated by light from a sub-stage condenser lens and lamp is usual. Two points in the object contained within the microscope slide do not then radiate entirely independently in phase. Abbe showed empirically that a good working rule for microscopes which takes this fact into account is to modify Equation 12.7 to:

$$h = \frac{0·5\lambda}{n \sin U} \qquad (12.8)$$

The unaided normal human eye is able to discern at the least distance of distinct vision (25 cm) detail corresponding to a linear separation between points of 0·01 cm (or 0·007 cm under ideal viewing conditions). Therefore, the microscope must provide a necessary magnifying power m_n given by:

$$m_n = \frac{0·01}{h} = \frac{0·01 \, (\text{N.A.})}{0·5\lambda}$$

For light of wavelength 5,500 Å, near that at which the eye has maximum sensitivity, the necessary magnifying power is, therefore,

$$m_n = \frac{0·01 \, (\text{N.A.})}{0·5 \times 5·5 \times 10^{-5}} = 360 \, (\text{N.A.})$$

If the linear magnification provided by the objective and eyepiece of the microscope exceeds this value considerably, no real advantage is gained because the eye is unable to perceive greater detail. Thus, with a numerical aperture (N.A.) of 1·35 for a good oil-immersion objective, the necessary magnification is $360 \times 1·35 = 490$ approx. A magnification in excess of this, up to about 650, can be an advantage for comfortable viewing but beyond this value the field observed will begin to become confused by too great a magnification of diffraction detail.

The limit of resolution of the microscope can only be made substantially less by reduction of the wavelength λ. A blue filter over the lamp which illuminates the sub-stage condenser provides modest assistance, and the insertion of this filter often clarifies intimate detail. The use of ultra-violet light is a possibility, provided that observation is by photography or by means of a fluorescent screen. Ultra-violet light microscopes are indeed available but expensive because of the demand for quartz optics; their advantage is in the contrast they provide for viewing certain biological specimens rather than in their resolution.

For extremely high resolution, the electron microscope is used. The de Broglie wavelength associated with electrons which have been accelerated through a p.d. of V volts is

$$\lambda = \frac{12 \cdot 26}{\sqrt{V}} \text{ Å}$$

At $V = 50{,}000$ V, λ is $5 \cdot 5 \times 10^{-2}$ Å. However, the object specimen must now be mounted in a vacuum, it must be transparent to the electrons, focusing of the electrons has to be by magnetic electron lenses, and viewing either by a fluorescent screen or photography. The severe aberrations of electron lenses for other than electron beams close to the axis means that the numerical aperture (N.A.) of the electron microscope cannot be made to exceed about 0·01. Thus, for an electron microscope at 50 kV, the limit of resolution is, from Equation 12.8, $(0 \cdot 5 \times 5 \cdot 5 \times 10^{-2})/0 \cdot 01 = 2 \cdot 8$ Å.

12.5 The Chromatic Resolving Power of a Prism

A prism of transparent material is illuminated by light from a collimator and, after refraction by the prism, the light is brought to the focal plane of a telescope objective where it is viewed by the telescope eyepiece. This is the well-known arrangement of the simple prism spectrometer, where the prism is set at minimum deviation.

Suppose the light from the source which illuminates the slit of the collimator contains components of wavelengths λ and $\lambda + \delta\lambda$ [Fig. 12.5(a)]. Provided that the collimator lens aperture is large enough, the rectangular face of the prism at which the light is incident forms the limiting aperture of the spectrometer. The spectrum observed in the telescope eyepiece will consist of two bright lines corresponding to the wavelengths λ and $\lambda + \delta\lambda$, and both these line images will be subject to Fraunhofer diffraction brought about by the slit source and the rectangular aperture formed by the prism face. The intensity variation in the light around the central maximum at each lens image centre will, therefore, be in accordance with Equation 11.2 and is shown graphically in Fig. 12.5(b) (in which the lines are assumed to be of the same intensity). The line images corresponding to wavelengths λ and $\lambda + \delta\lambda$ are said to be resolved if the Rayleigh criterion is maintained—i.e. if the central maximum in the intensity pattern for wavelength λ is not nearer to the central maximum for $\lambda + \delta\lambda$ than the first minimum for $\lambda + \delta\lambda$. The resolving power of the prism is decided by $\lambda/\delta\lambda$, where $\delta\lambda$ corresponds to the limit of resolution, at which the central maximum for λ coincides in position with the first minimum for $\lambda + \delta\lambda$.

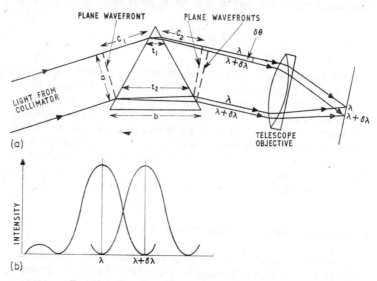

Fig. 12.5. The chromatic resolving power of a prism

From Equation 11.6, it is known that the angular separation of the first minimum from the central maximum at a wavelength λ is given by $\sin^{-1}(\lambda/a)$, where a is the width of the plane wave front incident upon the prism face. The spectral lines of wavelengths λ and $\lambda + \delta\lambda$ are, therefore, resolved if the angular separation between their line images is $\delta\theta$ [Fig. 12.5(a)] where

$$\delta\theta = \sin^{-1}\frac{\lambda}{a} = \frac{\lambda}{a} \qquad (12.9)$$

because $\delta\theta$ is small.

This angle of separation $\delta\theta$ will depend upon the angular dispersion $d\theta/d\lambda$ of the prism material. Now,

$$\frac{d\theta}{d\lambda} = \frac{d\theta}{dn}\frac{dn}{d\lambda} \qquad (12.10)$$

where n is the refractive index of the material; n is related to the wavelength λ by the Cauchy equation

$$n = A + \frac{B}{\lambda^2} + \frac{C}{\lambda^4}$$

in which A, B and C are constants, the last term C/λ^4 being negligibly small. Therefore,

$$\frac{dn}{d\lambda} = -\frac{2B}{\lambda^3} \qquad (12.11)$$

To determine $d\theta/dn$, consider the plane wavefronts shown in Fig 12.5(a). Optical paths between corresponding positions on the wavefronts must be the same. For the wavelength λ,

$$c_1 + nt_1 + c_2 = nt_2$$

For the wavelength $\lambda + \delta\lambda$,

$$c_1 + (n + \delta n)t_1 + c_2 + a\delta\theta = (n + \delta n)t_2$$

Subtraction gives:

$$a\delta\theta = (n + \delta n)(t_2 - t_1) - n(t_2 - t_1)$$

Therefore,

$$a\,d\theta = dn(t_2 - t_1) \qquad (12.12)$$

Substituting from Equations 12.11 and 12.12 into Equation 12.10 gives:

$$\frac{d\theta}{d\lambda} = -\frac{(t_2 - t_1)}{a}\frac{2B}{\lambda^3} \qquad (12.13)$$

From Equation 12.9, the smallest angle which can be resolved is $\delta\theta = \lambda/a$. Therefore,

$$\frac{d\theta}{d\lambda} = \frac{\lambda}{a\,d\lambda} = -\frac{(t_2 - t_1)}{a}\frac{2B}{\lambda^3}$$

The resolving power R of the prism is hence given by:

$$R = \frac{\lambda}{\delta\lambda} = \frac{(t_2 - t_1)\,2B}{\lambda^3}$$

This will be a maximum when $t_1 = 0$ and $t_2 = b$, the length of the base of the prism (i.e. when the whole of the prism face is illuminated by the incident light from the collimator). Therefore,

$$R_{max} = \frac{b\,2B}{\lambda^3}$$

For example, a prism of flint glass with a base 5 cm long has $2B/\lambda^3 = 1,000$ at $\lambda = 6,000$ Å; so $R_{max} = 5,000$. At a wavelength of 4,500 Å, the resolving power will be $(6/4\cdot5)^3 \times 5,000 = 12,000$ approx.

12.6 The Chromatic Resolving Power of a Plane Diffracting Grating

A plane diffraction grating has N rulings, and the centres of adjacent lines are separated by a distance e; so the full width of the rulings is $a = Ne$. Mounted on a spectrometer table, this grating receives incident light from a collimator of which the slit is illuminated by a source containing components of wavelengths λ and $\lambda + \delta\lambda$. Constructive interference results in principal maxima for normally incident light of wavelength λ at angles θ of the setting of the viewing telescope decided by Equation 11.19:

$$e \sin \theta = p\lambda$$

where p is an integer: 1 for the first order, 2 for the second, and so on. The dispersive power of the grating is $d\theta/d\lambda$. Now,

$$e \cos \theta \frac{d\theta}{d\lambda} = p$$

Therefore,

$$\frac{d\theta}{d\lambda} = \frac{p}{e \cos \theta} \qquad (12.14)$$

The dispersive power therefore increases with the order of the spectrum and with $1/e$, the number of rulings per centimetre.

If the components of wavelengths λ and $\lambda + \delta\lambda$ are to be just resolved then, in accordance with the Rayleigh criterion, the central maximum intensity in the line image due to light of wavelength λ must fall on the first minimum of the diffraction pattern due to light of wavelength $\lambda + \delta\lambda$. The angle $\delta\theta$ which separates these two images is given by Equation 11.7 as

$$\delta\theta = \frac{\lambda}{l}$$

where $\delta\theta$ is small, and l is the width of the wavefront in the parallel light leaving the grating to enter the telescope at the angle θ. It is seen from Fig. 12.6 that:

$$l = a \cos \theta = Ne \cos \theta$$

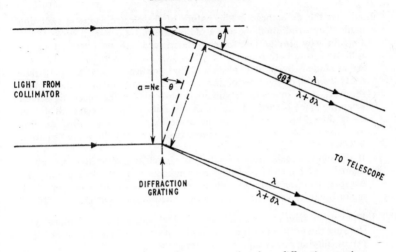

Fig. 12.6. The chromatic resolving power of a plane diffraction grating

Therefore,

$$\delta\theta = \frac{\lambda}{Ne \cos \theta}$$

Substituting for $\delta\theta$ from Equation 12.14 gives:

$$\frac{p \, \delta\lambda}{e \cos \theta} = \frac{\lambda}{Ne \cos \theta}$$

Hence, the resolving power R is given by:

$$R = \frac{\lambda}{\delta\lambda} = pN$$

The resolving power, therefore, increases with the total number N of lines used and the spectral order p.

Exercise 12

1. Describe how diffraction limits the ability of an instrument to produce resolved images of two point objects. How does the sensitivity of the method of radiation detection affect the practical limit of resolution? Electromagnetic waves of frequency 10,000 Mc/s reflected from distant aircraft are focused on to a detecting element by a scanning parabolic metal reflector 100 cm in diameter. What is the minimum separation of two aircraft at a range of 10 km if they are 'seen' separately by the receiving apparatus assuming that the Rayleigh criterion is obeyed?   (L. P.)

2. Discuss the differences in the nature of the electromagnetic radiation emitted by a sodium flame and a radio transmitter.

Explain why similar interference phenomena can be observed in the radiation from both sources.

An airplane is flying at a constant altitude vertically above the perpendicular bisector of the line joining a television transmitting station and a receiver. If an intensity maximum is observed on the receiving screen when the plane is vertically above this line, its distance from each station being then 5 km, how far will the plane have to travel before the next maximum is observed on the screen, if the wavelength of the radiation is 2 metres?

If the plane can be approximated to a circular diffracting aperture of diameter 12·2 metres projected perpendicularly to the line joining the plane to the transmitting station, estimate the length of travel over which major intensity variations might be expected in the receiver.   (L. P.)

3. Draw a ray diagram for an astronomical telescope in normal adjustment, showing the position of the exit pupil (eye ring). Explain the meaning of (a) magnifying power and (b) resolving power.

Calculate the magnifying power of an astronomical telescope consisting of two thin lenses, when the final image of a very distant object is formed at the least distance of distinct vision, 25·0 cm from the eyepiece. The focal lengths of the objective and eyepiece are 150·0 cm and 12·5 cm respectively.   (L. G.)

4. Discuss the diffraction pattern in the focal plane of a lens on which a parallel beam of monochromatic light falls, the lens aperture being restricted by a slit.

Explain what is meant by either the resolving power, or the limit of angular resolution of a telescope, and the dependence of this upon the aperture of the objective lens or mirror. Indicate the effect of this upon the design of modern telescopes for astronomical investigation. (L. Anc.)

5. Explain what is meant by (a) the magnifying power, and (b) the resolving power of a telescope and examine the factors that determine each of these quantities.

A telescope objective has a diameter of 5 cm. Calculate the greatest distance at which 5 parallel wires 1 mm apart could, in theory be distinguished when illuminated with the light of wavelength 5000 Å, and viewed through a telescope. What factors are likely to make the actual distance less than that calculated?   (L. Anc.)

6. Parallel light of wavelength 5461 Å falls normally on a converging lens of focal length 100 cm. Between the source and the lens is a fine slit of width 0·5 mm. Describe and explain the appearance and size of the pattern seen in the focal plane of the lens, deriving any formula required.

How could you investigate by experiment the variation of the resolving power of a telescope with aperture?   (L. G.)

7. A slit of width a is placed at right angles to a beam of monochromatic light and the diffracted light is brought to a focus by a converging lens.

Obtain an expression for the intensity along a line perpendicular to the length of the slit in the focal plane of the lens. At what point is the illumination zero?

A scale is graduated at intervals of $\frac{1}{2}$ mm. It is illuminated by light of wavelength 5461 Å and is focused by a telescope with its objective placed at a distance of 4 metres from the scale. Immediately in front of the objective a slit parallel to the scale rulings is adjusted in width until the rulings are just not distinguishable through the telescope. Explain why this is so and calculate the width of the slit. (L. Anc.)

8. *Either:* Give a descriptive account of the monochromatic aberrations of a coaxial lens system. Which of these are of particular importance in the performance of a high resolution microscope? Indicate how diffraction sets a limit to the degree of correction of aberration which it is profitable to achieve.

   Or: Derive from Snell's Law or from some equivalent principle, the relation between the transverse and angular magnifications of an optical system. Hence or otherwise find the relation between the focal lengths in the object and image spaces, assuming the refractive indices in these spaces to be different. Define the term nodal points and derive their positions. (L. G.)

9. Discuss either the Rayleigh or the Abbe theory of resolution by a microscope and obtain an expression for the resolving power of the objective. (L. P.)

10. Explain why the resolution of a microscope objective is limited, and discuss the factors on which this resolution depends.

    Describe the optical design of a high resolution, high magnification, microscope objective, emphasising the features essential for securing high resolution. (L. P.)

11. Describe briefly the factors in the design of a microscope objective which are important (a) in minimising its aberrations and (b) in achieving a high resolving power. Discuss in detail *either* (a) or (b). (L. P.)

12. Discuss the factors which affect the resolving power of an optical microscope.

    Review briefly the advantages and disadvantages of the electron microscope. (L. P.)

13. Two of the most important instruments in science are the telescope and the microscope. Write an account of *one* of these describing (a) its basic design and the features necessary to give optimum performance, (b) how it has been developed to cover wide spectral ranges and (c) some of its important scientific applications. (L. P.)

14. Describe the structure of a compound microscope explaining the functions of its optical parts. Explain what is meant by resolving power and depth of focus, and explain briefly how they are related to the focal length and the aperture of the objective lens. (L. Anc.)

15. *Either:* You may assume an expression of the form $\sin^2 \alpha/\alpha^2$ for the distribution of intensity in the Fraunhofer diffraction pattern from a single slit, as observed with monochromatic light incident normally. Hence or otherwise derive the corresponding distribution for a diffraction grating with six equal slits. If the slits are of width 0·02 cm and the separation of their centres is 0·05 cm describe in detail the diffraction pattern observed with light of wavelength 5461 Å in the focal plane of a lens of focal length 1 metre placed close to the grating. Estimate the positions and relative intensities of the principal maxima.

Or: Explain the meaning of (a) numerical aperture (NA) and (b) limit of resolution x of a microscope. Assuming that $x = (\lambda/2)(NA)$ when the object is illuminated by light of wavelength λ show that the greatest useful magnification is approximately 600(NA) when the image is viewed at the least distance of distinct vision 25 cm. Take λ as 5000 Å and the angular resolution of the eye to be 2 min of arc.   (L. G.)

16. Explain the Rayleigh criterion for the resolution of two point sources by an optical instrument.

(i) What is the smallest separation between point sources on the sun emitting radiation of 20 cm in wavelength which could be detected by a radio telescope with a dish diameter 40 metres? Express this as a fraction of the sun's diameter taking this as 10^6 miles and its distance from the earth as 93×10^6 miles.

(ii) What is the semi-angle of the beam it is necessary for the objective of an electron microscope to accept if it is to just resolve structures of 50 Å assuming the de Broglie wavelength of the electrons being used is 0·5 Å?   (L. P.)

17. Why is there a practical lower limit to the separation of radiating sources if they are to be distinguished as separate objects?

Two airplanes are flying side by side 100 m apart and are heading towards a radar installation operating at a wavelength of 3 cm. The receiving aerial is a parabolic dish of 1 m diameter. At what distance would it first be detected that there were two airplanes? Explain clearly any simplifying assumptions you make.

Calculate the smallest distance between two points which can be seen with a microscope, whose objective can be considered as a thin lens of 1 cm focal length and 1 cm diameter if the wavelength of the light used is 5000 Å.   (L. P.)

18. What is meant by chromatic resolving power? Derive an expression for the resolving power of a prism spectrometer. Calculate the resolving power of an equilateral glass prism of side 5 cm set at minimum deviation for a parallel beam of incident light. For the material of the prism

$$\frac{d\mu}{d\lambda} = 1,000 \text{ cm}^{-1}$$

for the light under consideration. State carefully the assumptions and approximations made throughout your answer.   (L. P.)

19. Find a relation between $d\theta/d\lambda$ of a prism used at minimum deviation, and $d\mu/d\lambda$ of the material of which the prism is made. Calculate the angular separation of the sodium lines 5890, 5896 Å using a 60° prism at minimum deviation if the refractive index of glass is 1·6043 at 5850 Å and 1·6031 at 5950 Å. (L. G.)

20. State what is meant by the term Fraunhofer diffraction and obtain an expression for the intensity distribution in the Fraunhofer diffraction pattern from a single slit.
A prism spectrometer is adjusted for parallel light and minimum deviation, and the width of the beam used is limited by a rectangular aperture. If p is the difference between the optical path lengths in glass travelled by the two extreme rays of the beam, show that the chromatic resolving power $\lambda/d\lambda$ of the prism is equal to $dp/d\lambda$. Describe and explain how to verify this relation experimentally. (L. G.)

21. Derive an expression for the smallest difference of wavelength between two spectral lines that can be resolved by a diffraction grating of width w cm, having n lines per cm, and used in the mth order. Hence find the approximate width of a grating with 1000 lines per cm, if it can resolve the sodium D lines in the third order but not in the first or second orders. Wavelength of one D line = 5890×10^{-5} cm, separation of D lines = 0.006×10^{-5} cm. (L. Anc.)

22. A plane transmission grating consists of transparent gaps of width a separated by opaque strips of width b and is illuminated by normally incident parallel monochromatic light. Obtain an expression (1) for the intensity distribution in the focal plane of a lens receiving the transmitted light, (2) for the chromatic resolving power. Explain what is meant by the latter term. (L. G.)

23. Derive an expression for the minimum difference of wavelength between two spectral lines that can be resolved by a diffraction grating of width w cm having n lines per cm and used in the mth order.
A plane diffraction grating deviates light falling normally on it through 28° in the first order. The grating is then immersed in a parallel-sided glass cell containing liquid of refractive index 1·40—the plane of the grating being parallel to the sides of the cell. What now is the deviation of the first order diffracted beam? (L. Anc.)

24. Give the theory of the plane diffraction grating, discussing the different roles played by diffraction and interference, and account for the existence of secondary maxima. Comment on the expression for its resolving power. (L. P.)

25. Discuss the concept and nature of 'resolving power' by considering (a) the microscope, (b) the telescope, (c) the spectroscope.
The diameter of the pupil of a human eye is 3 mm. At what distance in perfect viewing conditions would two sodium lamps, $\lambda = 6000$ Å, one metre apart, be just not resolved? (G, I. P.)

Answers to Numerical Exercises

Exercise 1, Page 51

1. 1·5 cm, convex towards P
2. 1·45
3. 30 cm on opposite side of lens
4. Principal and nodal points coincident at centre; focal points 1·52 cm from poles
5. Diverging lens of power 4 dioptres; 37·5 cm beyond lens
6. Principal plane between lenses, 1 cm from second lens; focal length $+1·5$ cm
7. (a) Outside system, $6\frac{2}{3}$ cm from each lens; (b) between lenses, $6\frac{2}{3}$ cm from respective lenses
8. $\frac{5}{3}$
9. $\frac{10}{13}$ cm outside sphere
10. Principal points at centre of 10 cm lens and at 10 cm outside 5 cm lens; focal points at centre of 5 cm lens; nodal points coincident with principal points
11. $z = 2$
12. Focal points at $66\frac{2}{3}$ cm in water and $25\frac{5}{9}$ cm in air; focal lengths $+25$ cm in air and $+66\frac{2}{3}$ cm in water; (hence positions of principal and nodal points)
13. Focal points at $9\frac{1}{6}$ cm in chamber and $1\frac{1}{3}$ cm from diverging lens between lenses; focal length of system $+3\frac{1}{3}$ cm; image between lenses, $2\frac{22}{27}$ cm from diverging lens; magnification $\frac{12}{27}$
14. Focal points at $17\frac{7}{9}$ cm from pole in water and 20 cm from pole in air; focal lengths $+20$ cm in air and $+26\frac{2}{3}$ cm in water
15. Diverging lens of focal length -25 cm
17. Focal point at $\frac{8}{11}$ cm from plane surface of hemisphere
18. Focal points at 40 cm on water side and 30 cm on air side
19. $\frac{3}{5}$ inches between lenses; image at $1\frac{10}{11}$ inches beyond second lens; magnification 15/11
20. With air between lenses, focal points between lenses at $1\frac{1}{4}$ cm from each lens, focal length of system $+6\frac{1}{4}$ cm; with liquid between lenses, focal

points at 8 cm outside the system; nodal points and principal points coincident in each case

21. Focal length $+100$ cm; focal points outside the system at 40 cm from concave lens and 220 cm from convex lens

Exercise 2, Page 83

1. 4 cm from lens on stop side and 3 cm in diameter
3. 75 cm
4. Common face radius 30 cm; radius of other face 20 cm
5. Principal planes at $l \bigg/ \left\{ n \left[2 - \dfrac{l(n-1)}{nr} \right] \right\}$ from poles; focal length

   $\dfrac{n-1}{r} \left[2 - \dfrac{l(n-1)}{nr} \right]$

6. (a) ± 5 cm; focal points at lens centres; (b) $+3\cdot75$ cm; focal points outside system, $1\cdot25$ cm from each lens
9. Common surface radius 3 cm; radius of other surface $1\frac{5}{7}$ cm

Exercise 3, Page 116

1. $+8\cdot45$ cm; $+4\cdot2$ cm and $+2\cdot4$ cm
2. $x' = \dfrac{xf^2}{F^2} + \dfrac{f}{F}(f+F)$
7. Focal length $+3\cdot2$ cm; focal points at $0\cdot8$ cm outside system; exit pupil nearly 1 cm from eyepiece and of diameter $0\cdot27$ cm
8. Focal length $+3$ cm; focal point at 1 cm from the 2 cm lens outside system and 3 cm from 6 cm lens inside system
9. As for Question 8
10. 6 cm beyond lens; $1\cdot75$ cm
13. Focal length $+2\frac{1}{4}$ cm; focal planes $\frac{3}{4}$ cm outside each lens
14. Focal length $+2\frac{2}{3}$ cm; focal points $\frac{2}{3}$ cm beyond 2 cm lens and virtual 4 cm beyond 4 cm lens; Ramsden disc at $1\cdot16$ cm from eye lens and of diameter $0\cdot537$ cm; magnification $5\frac{5}{8}$

Exercise 4, Page 135

2. (a) $1°24'$; (b) $1\cdot222$ cm
4. $1\cdot67$
6. 2×10^{-5}
7. $1\cdot624$ to $1\cdot280$
11. $43°26\frac{1}{2}'$

Exercise 5, Page 165

4. $1\cdot66 \times 10^{10}$ cm per sec
7. $1\cdot73 \times 10^{10}$ cm s^{-1}

Exercise 6, Page 208

7. 9% approx.
9. 4,833 Å

Exercise 7, Page 234

2. 1.78×10^{10} cm sec^{-1}
3. 1.66×10^{10} cm sec^{-1}
6. 5 km sec^{-1}

Exercise 8, Page 245

1. 8.2×10^{-4} cm
4. $I = 4a^2 \cos^2 \left[\dfrac{\pi}{2} \left(1 + \dfrac{4h \sin \alpha}{\lambda} \right) \right]$

Exercise 9, Page 259

1. n_R is 10% greater than n_B
2. 90 Å
4. 1.06×10^{-4} rad
5. 2.95×10^{-4} rad
7. 0.768 cm; 5156 Å

Exercise 10, Page 280

4. $2v/\lambda$
5. 2.73×10^{-10} rad
10. 5.79 Å

Exercise 11, Page 312

3. 2.78
5. 4th order 7,500 Å; 5th order 6,000 Å; 6th order 5,000 Å; 7th order 4,286 Å
6. Bright bands visible at intervals of 9.38′
8. Bright bands visible at intervals of 34′
9. 1%
10. 1.000292
12. 0.038%
13. 0.33 mmHg
15. 5×10^{-8} rad
16. 16.4 mmHg min^{-1}
19. $a = 3.3b$ approx.
23. 5th order 6546 Å; 7th order 4676 Å; 8th order 4091 Å
24. 2nd order 5875 Å
25. 4°46′ and 36°56′

Exercise 12, Page 329

1. 366 m
2. 100 metres; 2,000 metres
3. 18
5. 100 m
6. First dark band 0·109 cm from the axis
7. 0·437 cm

15.

Order of interference	0	1	2	3	4	5
Relative intensity	1·00	0·57	0·05	0·025	0·035	0
Position (min of arc)	0	3·5	7	10·5	14	17·5
Position in focal plane of lens, from axis (cm)	0	0·11	0·22	0·33	0·44	0·55

16. (i) 0·57; (ii) 21′
17. 2740 m; $6·8 \times 10^{-5}$ cm
18. 5,000
19. 0·4′
21. 0·33 cm
23. 28°
25. 4100 m

Index